Vehicular Communications for Smart Cars

Vehicular Communications for Smart Cars

Protocols, Applications and Security Concerns

Niaz Chowdhury and Lewis Mackenzie

CRC Press
Taylor & Francis Group
Boca Raton London New York

CRC Press is an imprint of the
Taylor & Francis Group, an **informa** business

First edition published 2022
by CRC Press
6000 Broken Sound Parkway NW, Suite 300, Boca Raton, FL 33487-2742

and by CRC Press
2 Park Square, Milton Park, Abingdon, Oxon, OX14 4RN

ISBN: 978-0-367-45744-0 (hbk)
ISBN: 978-1-032-10872-8 (pbk)
ISBN: 978-1-315-11090-5 (ebk)

DOI: 10.1201/9781315110905

Typeset in Sabon
by KnowledgeWorks Global Ltd.

Contents

Preface

The use of vehicles, in the widest sense of the word, has a long history in human society. The availability of wireless radio communication is somewhat more recent but, even so, commercial vehicles such as ships, airlines, and railways have been using the technology for over a century. For example, when the Titanic issued its distress signal using radio telegraphy in 1912, the technology was in wide use in the North Atlantic despite having been available for only a decade. It is perhaps surprising that, where road vehicles are concerned, it has taken so long to form any kind of widely deployed Vehicle to Vehicle (V2V) communication system, and even in today's world connected "smart cars" still belong mostly in textbooks. However, we are of the view that the technological, commercial, and political conditions are now in place to transform this situation very rapidly in the near future.

A V2V system can offer a wide range of powerful applications to improve the driving experience. The safety applications that such a technology would enable could undoubtedly greatly reduce the number of accidents. At the same time, value-added services such as digital maps, traffic information, and better routing could enhance the driving experience and shorten travel times. Entertainment applications such as web browsing, reading, gaming, movies, and music could make the journey more pleasant for passengers. These applications require appropriately designed underlying protocols and supporting technologies to make them available securely in a highly mobile environment.

This book offers eight chapters that examine the main underlying elements that might make a regular vehicle smart. Nevertheless, it would be impossible to cover a broad topic like V2V communication with appropriate depth in a single book of this nature and emphasis is given, rather, to addressing the key issues that arise in the area. The use of the Internet of Things (IoT), big data, and broadcasting techniques is examined in various chapters, emphasizing security, warning dissemination, the communication networks available, and other enabling technologies such as blockchain.

Chapter 1 reviews the IoT-enabled use of visible light optical camera communication in smart cars. Chapters 2 and 3 describe the accident warning system and its use of broadcasting in message dissemination, respectively. Chapter 4 reviews a business aspect in the mix by investigating the uses of big data in smart city transportation to accelerate business growth. Chapter 5 takes a futuristic approach to secure smart vehicles using blockchain technology in the quantum era. The remaining three chapters, 6, 7, and 8, investigate security issues in autonomous vehicles, communication networks, and accident warning systems.

Although doctoral students and early career researchers are the primary audience and the material is pitched at that level, senior researchers and academics should also appreciate its content. The author list of the book contains a mixture of academics, consultants, and scientists, allowing it to present a portfolio of contributions. The editors applaud their participation and hope to see these efforts contribute to making the smart car concept a widespread reality in the near term.

Editors

Dr Niaz Chowdhury is a research associate at the Open University (OU). Following his doctoral study from the University of Glasgow, he joined the OU as a postdoctoral researcher in 2015, where he first worked in the smart city project MK:Smart. In this project, his role involved developing IoT-based holistic energy management solutions for consumers with electric vehicles. Dr Chowdhury then joined OU's Strategic Data Science project, where he researched blockchain-based solutions for bakeries and restaurants to manage regulatory compliance in a trusted environment. More recently, he has been working in the EU-funded project, QualiChain, to develop a blockchain-based decentralized platform for storing, sharing, and verifying education and employment qualifications. He is also involved in another EU-funded project, DEL4ALL, that aims to assess the challenges and opportunities offered by the increasing adoption of digital technologies, such as blockchain and artificial intelligence (AI). Dr Chowdhury authored a book entitled 'Inside Blockchain, Bitcoin, and Cryptocurrencies' published by Taylor & Francis in 2019 and also has been serving as a series editor for the Smart Technology series for the same publisher. He has authored and co-authored many peer-reviewed articles in reputed journals, conferences and workshops on the topics of blockchain technology, decentralization, and privacy issues. Dr Chowdhury was a Scottish ORS Award recipient at the University of Glasgow, Scotland, a Govt. of Ireland Scholar at Trinity College Dublin, Ireland, and a Gold Medalist at East West University, Bangladesh.

Dr Lewis M. Mackenzie is a senior lecturer in Computing Science at the University of Glasgow. His research interests are in machine architectures, the performance modeling of communication systems, Internet of Things, vehicular networks, theory of computation and usable security. Dr Mackenzie's recently published work has addressed the modeling of traffic patterns in a variety of network scenarios from regular wormhole-switched multi-computer interconnects to mobile ad hoc wireless networks (MANETs), routing in RPL-based IoT networks, clustering algorithms in UAV formations and V2V safety applications.

Contributors

Geetika Aggarwal completed her PhD in January 2020 from Northumbria University, UK, in the field of Electronics and Communication Engineering, her research was focused on design a wireless system on chip for healthcare using Visible Light Optical Camera Communication. Presently, She is working as RA at NTU, with research areas including Mobile Computing, Optical Camera Communication, Internet of Things, Wireless Communication, Computer Vision, Image Processing, 5G and beyond.

Bannishikha Banerjee is Research Scholar at PP Savani University, Surat, India. She has more than seven research publications in reputed journals. She achieved gold medal in her masters of engineering from Gujarat Technological University. She has more than 4 years of teaching and research experience. Her area of expertise is Blockchain, Machine Learning, Smart Vehicles, IoT, and Cryptography.

Md Aminul Islam completed his BSc in Engineering from the Bangladesh University of Engineering and Technology, Dhaka, Bangladesh. He also completed his MSc from the Institute of Disaster Management and Vulnerability in University of Dhaka, Bangladesh. Currently he is pursuing his post-graduate research from the Oxford Brookes University in the United Kingdom under the faculty of Technology, Environment and Design for Advanced Computer Science. He has written several books on mathematics and physics for high school students published by Jupiter, Lecture, Anindya, and Ittadi Publication. He also worked as an IT support engineer at ICT division and data center of Standard Bank Limited, Dhaka, Bangladesh.

Ashish Jani is Associate Professor at PP Savani University, Surat, India. He completed his post-doctoral fellowship from Florida Atlantic University, United States. He has more than 15 years of teaching and research experience. He is a certified data scientist. He has more than 15 research publications in several journals of repute. He has published four books. He received grant of INR 7,50,000 for his research work from GUJCOST. His area of interest is Computer Vision, IoT, Machine Learning, and Deep Learning.

Pooja Jha received her PhD degree in Technology from BIT Mesra, Ranchi, India, in 2019 and completed her master's degree from BIT Mesra, Ranchi, India. She is currently working as an assistant professor in the Department of CS-IT, Amity University Jharkhand, India, since 2017. She has more than 14 years of teaching experience. She has contributed more than 11 research papers. She has published papers in many reputed journals like International Journal of System Assurance Engineering and Management, International Journal of Computer Applications in Technology, Journal of Theoretical and Applied Information Technology, etc. She has been a reviewer of International Journal of Software Innovation (IJSI) and an active member of International Association of Engineers (IAENG). She has handled many university students with their scholarly research papers. Recently, she has been assigned as a reviewer of IEEE Madras Section International Conference, MASCON 2021. Her research interests mainly focus on Software metrics, Artificial Intelligence, Machine Learning, and Soft Computing.

Md Khaladun Nabi, Independent Research Analyst. I consider myself as an independent researcher in the field of business where there are so many facts of business (e.g., business management, human resource, business analytics, artificial intelligence (AI) in business, business entrepreneurship, machine learning in business, economics, etc.) that are inextricably linked together that I do not think a researcher in this field can be associated with just one area. I became interested in research in 2010 while doing my master's dissertation at Birmingham City University, United Kingdom, where my dissertation was on cultural diversification and recruiting strategies in multinational organizations in advanced and evolving countries. From there, I continued my research in AI in HR strategies, the impact of AI in emerging economies workforce, talent acquisition through AI, etc., and published articles in international journals. I have cooperated with community members, non-profits, academics, and business representatives on several projects in the United Kingdom and Bangladesh as well. This variety of experience in the area of modern business and technology contributed to my ability to perform the level of editing, writing, and research required in my publishing contracts. I have focused on publishing books, articles in journals for the last several years as an independent research analyst.

Ranjana Lakshmi Patel received her master's in Cybersecurity from Northumbria University, UK. In 2015, she completed her bachelor's degree from Meenakshi College of Engineering in Computer Science. She has publications in several research papers in various international journals and conferences. She previously worked as a Software Testing Engineer in Equiniti.

Sarah Al Qahtani pursued studies in Electrical Engineering at the American University of the Middle East in Kuwait and has joined the Robotics Club, IEEE, IET, and many other international organizations as an active member and volunteer while completing her bachelor's. She has acquired significant working experience with prominent international companies including Microsoft and Honeywell, demonstrating competent skills in the workforce. To further advance her career, she sought a master's degree in computing science from the Oxford Brookes University in the United Kingdom.

Niraj D Shah is Dean of the School of Engineering in PP Savani University, Surat, India. Dr Shah has an experience of 22 years, including academic, administrative, industrial, and research experience. He has several publications, authored books, guided Doctoral and PG students, and has been honored with numerous prestigious awards. Dr Shah has been consulted in various projects. He has been invited to deliver several key note/plenary/expert lectures and to present scientific papers in several international and national conferences in India and abroad.

Rejwan Bin Sulaiman is currently pursuing his PhD degree in Computer Science from the University of Bedfordshire, UK. He graduated from the same university with an MSc in Computer Science in 2019 and a BSc from Wrexham Glyndwr University in 2018. Besides his academic qualifications, he also holds professional certifications like CEH, CCNA, AWS. His research interest focuses around Cybersecurity, Artificial Intelligent, and Machine Learning. He regularly participates in scientific meetups and has presented his work in various conferences and journals. Most recently he is working as a teaching assistant under his supervisor at the University of Bedfordshire.

Chapter 1

A review of Internet of Things (IoT) using visible light optical camera communication in smart cars

Geetika Aggarwal

Nottingham Trent University, Nottingham, United Kingdom

CONTENTS

1.1 INTRODUCTION

The exponential increase in data rate usage by end users is constantly increasing the demand of capacity of wireless protocols. Optical wireless communication (OWC) offers a huge, unregulated bandwidth spectrum that is unoccupied and can be utilized in communication to alleviate the radio frequency (RF) spectrum crunch. Furthermore, in the past decade, OWC has diverted the attention of researchers to meet the growing data traffic demand and to offload the congested RF networks.[1–6] The LED market is growing at a fast pace, and as a result visible light communication (VLC) systems deploying LEDs are increasingly being used in numerous applications. Considering the marvelous improvement in smart devices in recent years, most of these devices are furnished with LED lights and cameras. This opens

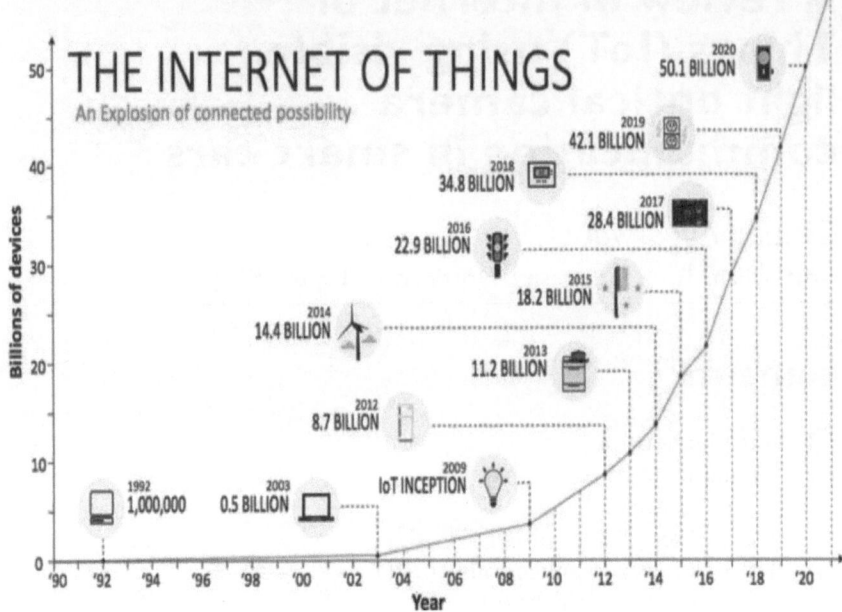

Figure 1.1 Exponential growth of IoT-connected devices.[15]

a probability of VLC execution for these gadgets that utilization a camera as the handset pair, without a need to uphold extra equipment adjustments.[7–10] The Internet of Things (IoT) is an outstanding evolution that enables communication between numerous devices through sensors, actuators, embedded systems, and various other technologies via the Internet. Figure 1.1 shows the exponential growth of IoT and interconnected devices.[11–14]

Figure 1.2 illustrates the scenario of a connection between a physical device, a vehicle in this case, and the Internet, where the vehicle has multiple devices, such as sensors and actuators, that are used for communication through the Internet, resulting in IoT. This chapter explores the area of smart cars with IoT using camera communication. The rest of the chapter is divided into the following sections: Section 1.2 describes the fundamental theory; Section 1.3 discusses the potential applications of IoT in smart cars with VL-OCC; and Section 1.4 explores future research directions.

1.2 FUNDAMENTAL THEORY

1.2.1 History

The term *Internet of Things* (IoT) was coined in 1999 by Kevin Aston at Procter & Gamble for his research. The IoT is directly related to evolution in communication systems.[16–20] The advancement in communications

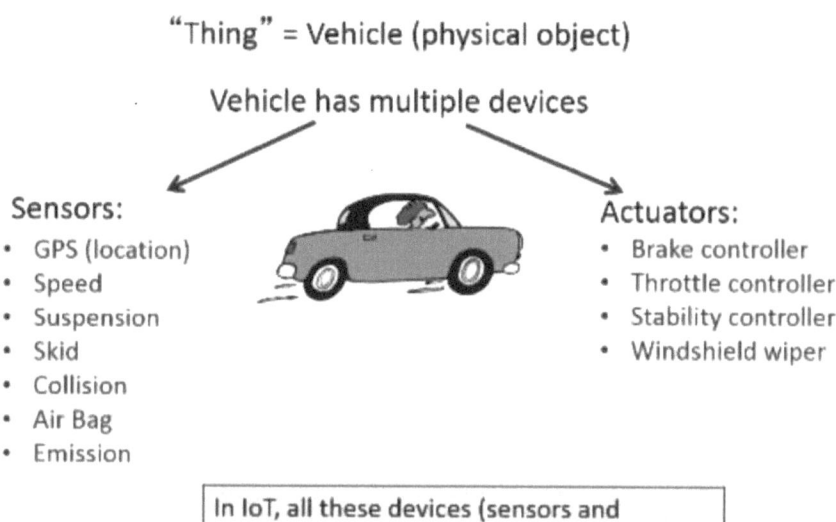

"Thing" = Vehicle (physical object)

Vehicle has multiple devices

Sensors:
- GPS (location)
- Speed
- Suspension
- Skid
- Collision
- Air Bag
- Emission

Actuators:
- Brake controller
- Throttle controller
- Stability controller
- Windshield wiper

In IoT, all these devices (sensors and actuators) can be accessed via the Internet!

Figure 1.2 Physical device such as car connected through IoT.[15]

technology, the increase in data transfer, and the demand for interconnected devices have resulted in gradual increase of use of the Internet to connect the devices, thus resulting in the IoT. The basic idea of IoT is that virtually every physical thing in the world can become a computer that is connected to the Internet, resulting in ubiquitous smart computers or smart devices.[21–27] For instance, a consumer good could be considered to be "smart" when tagged with a visual code such as a bar code or equipped with a time-temperature indicator that, say, a mobile phone can use to derive and communicate the product's state of quality, dynamic carbon footprint, effect on diabetics, or origin.[28, 29] Certainly, the boundary is blurring between smart things, which autonomously can derive and transform to different states and communicate these states seamlessly with their surroundings, and not-so-smart things, which only have a single status and are not very active in communicating it.[30–35]

1.2.2 Architecture of IoT

The IoT architecture comprises five different layers: perception layer, network layer, middleware layer, application layer, and business layer. The perception layer is the bottommost layer and is composed of physical devices such as sensors and actuators that are responsible for collecting the information and transferring it to the network layer.[36–41] The transmission of information to the information processing system is done by the network layer. This data/information transfer is possible or done using wired or wireless communication protocols and so forth. After the network layer, the next layer is middleware layer, whose task is to process the information received from the network layer and help in aiding the decisions that can be further used by the application layer for global device management.[42–48] The topmost layer in the IoT architecture is the business layer, which is responsible for the overall IoT system, connectivity, applications, and services.

Besides the layered framework, the IoT system consists of several functional blocks, shown in Figure 1.3, that support several IoT activities, such as the sensing mechanism, authentication and identification, control, and management.[49–55]

These functional blocks are responsible for input/output operations, processing data, and storing data. For optimum performance of an overall IoT system, all these functional blocks are interrelated. The key attribute of IoT architecture is scalability; the architecture must designed in such a way that it is scalable and is able to provide user-friendly applications. Figure 1.4 shows the modern architecture of IoT, which is stage 4 architecture.

Stage 1 comprises real-world elements such as sensors and actuators, for interconnectivity by detecting the signal and data transfer, followed with further analysis of data.[55–62] Also, actuators are used in temperature control, turning off lights and music, and so on. Hence, stage 1 is focused on collecting real-world data that could be useful for further analysis.[56, 57]

Stage 2 is responsible for collaboration through gateways and data acquisition systems with sensors and actuators. In stage 2 the data collected or generated from stage 1 is aggregated and optimized in a structured way suitable for processing, which is then passed to stage 3, comprising edge

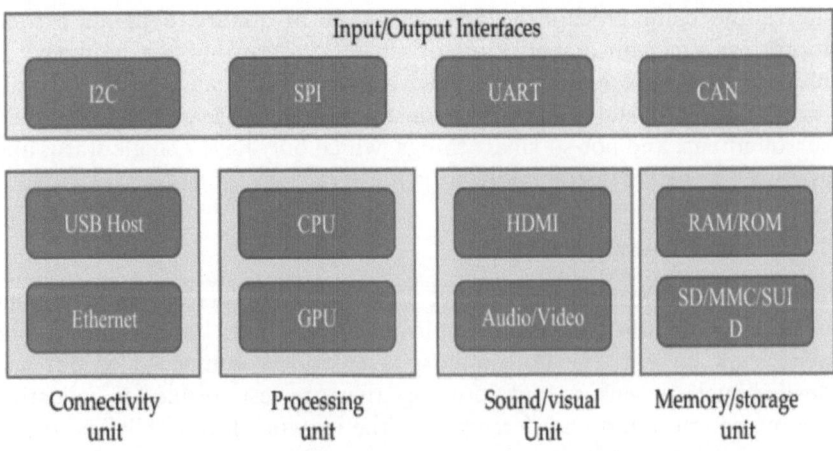

Figure 1.3 Functional blocks of IoT architecture.[55]

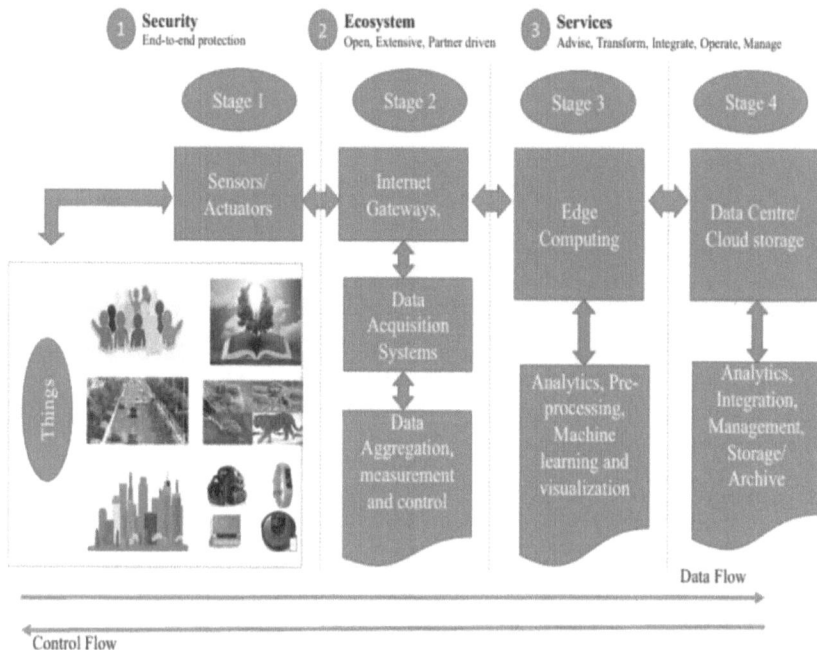

Figure 1.4 Different stages of IoT architecture.[55]

computing. *Edge computing* can be defined as an open architecture in distributed fashion that allows use of IoT technologies and massive computing power from different locations worldwide. Edge computing is a very powerful approach for streaming data processing and thus is suitable for IoT systems.

In stage 3, edge computing technologies deal with massive amounts of data and provide various functionalities such as visualization, integration of data from other sources, analysis using machine learning (ML) methods, etc. Stage 4 consists of several essential activities such as in-depth processing and analysis, sending feedback to improve the precision and accuracy of the entire system. Everything in stage 4 is performed on cloud servers or in a data center. A "big data" framework such as Hadoop or Spark may be utilized to handle this large quantity of streaming data, and ML approaches can be used to develop better prediction models that could help in producing a more accurate and reliable IoT system to meet the ever-increasing demands.[55, 63–67]

1.2.3 Potential applications of IoT

IoT has been leveraged in numerous applications, some of which are listed in Figure 1.5 and discussed in this section.

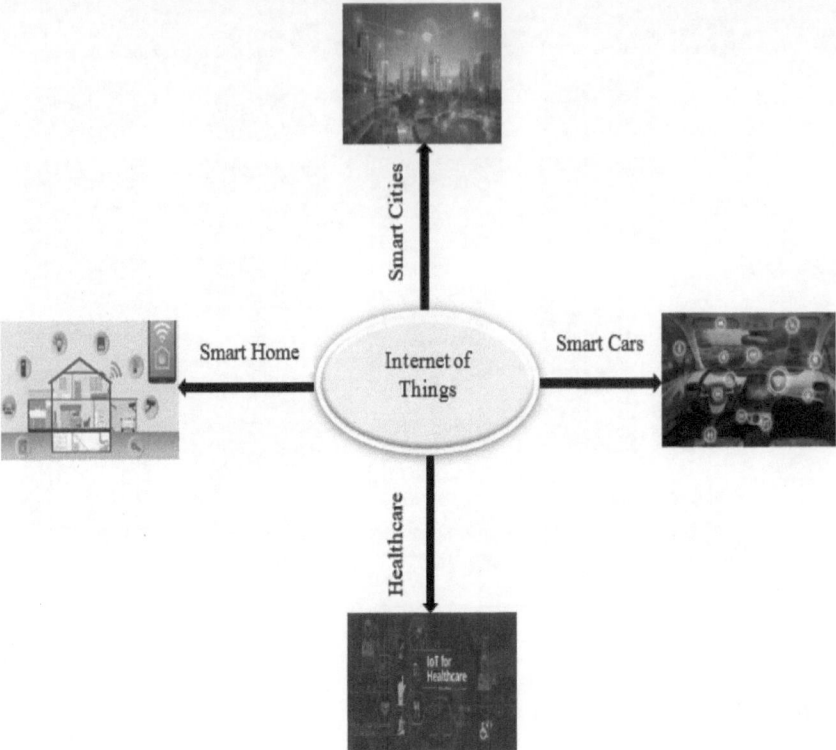

Figure 1.5 Potential applications of Internet of Things.

1.2.3.1 Healthcare

The reliance of healthcare on IoT is increasing by the day to improve access to care, increase the quality of care, and, most importantly, reduce the cost of care.[69, 70] IoT devices can provide remote health monitoring and emergency warning systems. These fitness and health monitoring devices can vary from blood pressure monitors to advanced devices that are capable of monitoring crucial implants such as a pacemaker. For example, sensors can be used to monitor the health of elderly persons, with an emergency warning mechanism for alerting the nearest hospital in case of an emergency. The emerging wireless sensor technology provides the capability to continuously sense, process, and transmit the required signals to a control station.[71, 72]

1.2.3.2 Smart home

The term *smart home* refers to connecting devices such as an HVAC system, lights, locks, microwaves, machines, and so forth through the Internet. For example, automated alerts could help in switching on/off lights and thus help in saving electricity and power consumption. For the elderly, smart homes could be helpful in supporting health and well-being too by interconnecting the IoT devices through the Internet without any extra or additional effort.[73–77]

1.2.3.3 Smart cities

With the increase in urbanization and the growing population in cities, the necessity to monitor and manage power consumption, air quality, noise, traffic, healthcare, and infrastructure have resulted in the move toward the interconnected devices through internet deploying IoT in smart cities. For example, people's lives would become much easier and more productive through IoT in smart cities by knowing the information such as free parking slot, thus saving precious time in search. Hence, IoT has immense potential in transforming cities into smart cities that improve citizens' daily life.[78–83]

1.2.3.4 Smart cars

IoT is a disruptive technology where the cyber world meets the physical world. It is autonomous communication between inanimate objects[84] to benefit human beings. IoT encompasses all technologies in social, mobile, analytics, and cloud (SMAC). The automotive industry is on course to a disruptive transformation using developments around smarter vehicles and related infrastructure. IoT is at the heart of this digital transformation in the auto sector. IoT connects people, machines, vehicles, auto parts, and services to streamline the flow of data, enable real-time decisions, and improve automotive experiences.[85] Leading automotive manufacturers, suppliers, and dealers have started investing heavily in IoT and are gaining returns in the form of ultra-efficient inventory management, real-time promotions that grow sales, reduced operational expenses, and increase in revenue. They are beginning to change their business processes and to recognize that, in time, IoT will touch every area of automotive operations and customer engagement.[86–90]

The preceding potential applications of IoT were just a small sampling of the numerous applications that are possible. The next section discusses the application and use of IoT in cars.

1.3 IoT DEPLOYING VL-OCC IN SMART CARS

The widespread use of smart devices equipped with LED-based screen lighting, flashlight, and quality cameras offers the opportunity to establish VLC links, where the flashlight and the camera can be used as a transceiver without the need for additional hardware. In contrast to the single photodiode (PD) based visible light communication (VLC) systems, a camera-based receiver (Rx) in optical camera communication (OCC), which is composed of an imaging lens and an image sensor (IS), has many unique features,

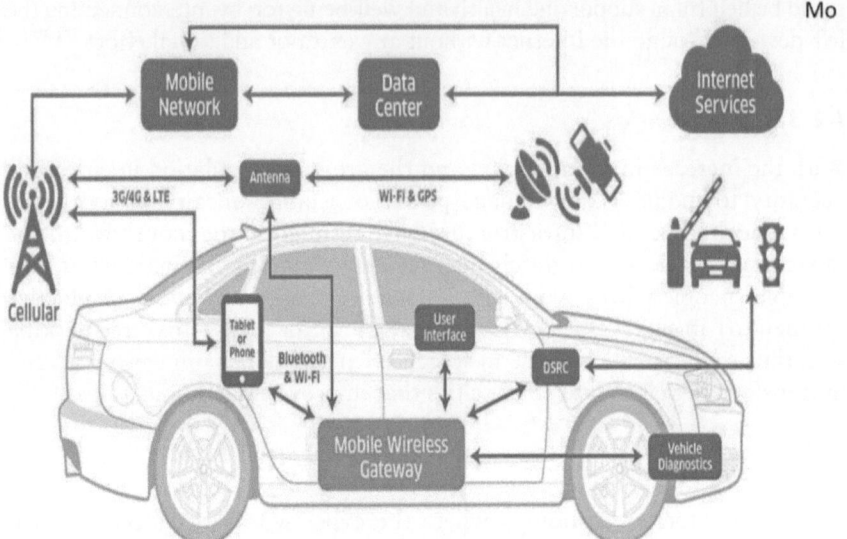

Figure 1.6 Description of a car's connectivity in IoT.[92]

including a wide field of view (FoV) due to the PD array as well as spatial and wavelength separation of the light beams.[92–94] Figure 1.6 shows the description of a car's connectivity in IoT.

VL-OCC can help in smart cars mainly for transmission of safety warnings and traffic information, which can be effective for avoiding accidents and traffic congestion.[7] Cameras (i.e., a matrix of PDs) can be employed for object and people detection as well as collision warning, enhanced driving safety, range estimations, and data communications in vehicle-to-vehicle (V2V) and vehicle-to-infrastructure (V2I) communications.[6, 8, 9]

Moreover, with the introduction of autonomous vehicles, localization and communications between vehicles and the surrounding environment (i.e., the roadside infrastructure, etc.) has become an essential part of future smart cities.[11] Within this context, intelligent transport systems (ITS) have been proposed to support improving the traffic flow, reducing pollution and road accidents, reducing energy usage, and improving economic productivity and the quality of life.[12–14] An ITS enables sharing of information between vehicles and/or vehicles and road infrastructures. The current ITS technologies are also known as vehicular ad hoc networks (VANET), V2V, or V2I communications, and are RF-based under the name of dedicated short-range communications (DSRC).[15] In an ITS, vehicles can communicate with each other without the use of cellular wireless base stations. The main features of VANET are safety, improved road usage efficiency, and providing in-vehicle information/entertainment.[16–18]

In addition, cameras can be used for multiple purposes such as vision, surveillance, biometric facial recognition,[38] scene change detection,[39] augmented reality,[40] and positioning.[41] Nowadays, new vehicles come with at least two front cameras and one rear camera for the purpose of people and object detection, smart lighting (i.e., estimating the position of oncoming vehicles using the camera and adjusting the front lights' intensity and profiles to avoid dazzling),[42] parking, ranging, and so forth.

The following are examples of scenarios in which VLC-OCC IoT in smart cars can be beneficial:

a. Vehicles approaching a road junction or a crossroad from different directions, where there is no line of sight (LOS) link between them.
b. Vehicles traveling in the same direction that are not in the FOV of either the front or rear camera of each other. Hence, the need for vehicles to send information for lane changing.
c. High-profile vehicles blocking the LOS link between the vehicle and roadside lights.
d. In urban areas, tall trees, especially in the spring and summer, may block the LOS link between the streetlights and the vehicles.[8] Hence, information about vehicles approaching a junction and cyclists or pedestrians crossing the road is achievable via a NLOS link.
e. Tall streetlights may not be in the FOV of the vehicles, especially at roundabouts and in urban areas, and therefore unable to transmit the information about the traffic conditions, emergency vehicles, etc.
f. In heavy traffic conditions, where the LOS between cars might not be available, reflections from the road surface may be the option to establish communications.

The communication of smart cars using camera and LED lights deploying VL-OCC has an immense potential in reducing road accidents, locating vehicles parking spaces, and many other applications.

1.4 FUTURE RESEARCH DIRECTIONS

This section describes some possible research directions in the field of IoT with smart cars. IoT involves transfer of data among different technologies through embedded systems, which gives rise to several open areas of research to be considered. The following examples represent just a few of the many possibilities.

1.4.1 Security

As vehicles become more connected, networks that previously did not extend beyond the vehicle itself are now being connected to the Internet and other external networks. This has led to concerns about attacks on vehicles from these external networks, raising the issue of how to reduce security risks.

One of the most important and challenging issues regarding IoT is security and privacy, due to the various threats, cyber-attacks, risks, and vulnerabilities.[101] The issues that give rise to device-level privacy are insufficient authorization and authentication, insecure software, firmware and web interface, and poor transport layer encryption.[102] Addressing security and privacy issues is very important to develop confidence in IoT systems with respect to various aspects.[103] Security mechanisms must be embedded at every layer of IoT architecture to prevent security threats and attacks.[104–107] Therefore, maintaining the authorization and authentication over a secure network is required to establish communication between trusted parties.[46] Another issue is the different privacy policies for different objects communicating within the IoT system. Each object should be able to verify the privacy policies of other objects in an IoT system before transmitting data.

1.4.2 Intelligence

Ambient intelligence and autonomous control are not part of the original concept of the IoT. Ambient intelligence and autonomous control do not necessarily require Internet structures, either. However, there is a shift in research to integrate the concepts of IoT and autonomous control, with initial outcomes toward this direction considering objects as the driving force for autonomous IoT.[108–112] In the future, IoT may be a nondeterministic and open network in which auto-organized or intelligent entities (web services, SOA components, etc.), and virtual objects (avatars) will be interoperable and able to act independently (pursuing their own objectives or shared ones) depending on the context, circumstances, or environments. Autonomous behavior through the collection and interpretation of context information as well as the objects' ability to detect changes in the environment such as faults affecting sensors and introduce suitable mitigation measures constitute a major research trend, clearly needed to provide credibility to the IoT technology. Modern IoT products and solutions in the marketplace use a variety of different technologies to support such context-aware automation, but more sophisticated forms of intelligence are required to permit sensor units to be deployed in real environments.[109–112]

1.4.3 Power consumption

With the increasing demand of data rate among users, numerous devices interconnected using the Internet, there is a need to keep a check on power consumption by developing devices that require minimum power, thus moving toward miniaturization by reducing the size and weight, which will reduce the power consumption of the specific device. Even with improved batteries and green energy sources like solar and wind, just meeting the data rate demand is becoming difficult, which is another future research direction.[108–114]

1.5 CONCLUSION

This chapter listed the scope of IoT in different sectors, including in smart cars using VL-OCC. IoT with connected devices through Internet specifically cars have an immense potential to minimize road accidents, improve traffic flow, reduce energy consumption etc. With the deployment of existing devices such as LED lights and cameras in cars, VL-OCC can be used for data communication to transmit data over the Internet wirelessly, thus reducing the overall cost of the system. IoT using VL-OCC in smart cars can be helpful in detecting road accidents, providing more accurate weather forecasts, finding available parking spaces, vehicle-to-vehicle communication, vehicle-to-infrastructure communication, and driver safety monitoring. This chapter also discussed future research directions that IoT researchers can engage in to get optimum solutions to the marker sooner.

REFERENCES

1. Khajenasiri I, Estebsari A, Verhelst M, Gielen G. A review on Internet of Things for intelligent energy control in buildings for smart city applications. Energy Procedia. 2017;111:770–9.
2. Internet of Things. http://www.ti.com/technologies/internet-of-things/overview. html. Accessed 01 Apr 2019.
3. Liu T, Yuan R, Chang H. Research on the Internet of Things in the automotive industry. In: ICMeCG 2012 international conference on management of e-commerce and e-Government, Beijing, China. 20–21 Oct 2012. p. 230–3.
4. Alavi AH, Jiao P, Buttlar WG, Lajnef N. Internet of Things-enabled smart cities: state-of-the-art and future trends. Measurement. 2018;129: 589–606.
5. Olivier F, Carlos G, Florent N. New security architecture for IoT network. In: International workshop on big data and data mining challenges on IoT and pervasive systems (BigD2M 2015), procedia computer science, vol. 52; 2015. p. 1028–33.
6. Noura M, Atiquazzaman M, Gaedke M. Interoperability in Internet of Things: taxonomies and open challenges. Mob Netw Appl. 2019;24(3):796–809.
7. Al-Fuqaha A, Guizani M, Mohammadi M, Aledhari M, Ayyash M. Internet of Things: a survey, on enabling technologies, protocols, and applications. IEEE Commun Surv Tutor. 2015;17(June):2347–76.
8. Palattella MR, Dohler M, Grieco A, Rizzo G, Torsner J, Engel T, Ladid L. Internet of Things in the 5G era: enablers, architecture and business models. IEEE J Sel Areas Commun. 2016;34(3):510–27.
9. 57. Fafoutis X, et al. A residential maintenance-free long-term activity monitoring system for healthcare applications. EURASIP J Wireless Commun Netw. 2016. https://doi.org/10.1186/s13638-016-0534-3.
10. Park E, Pobil AP, Kwon SJ. The role of Internet of Things (IoT) in smart cities: technology roadmap-oriented approaches. Sustainability. 2018.

11. Lee C, Yeung C, Cheng M. Research on IoT based cyber physical system for industrial big data analytics. In: 2015 IEEE international conference on industrial engineering and engineering management (IEEM). New York: IEEE; 2015. p. 1855–9.

12. Rizwan P, Suresh K, Babu MR. Real-time smart traffic management system for smart cities by using internet of things and big data. In: International conference on emerging techno-logical trends (ICETT). New York: IEEE; 2016. p. 1–7.

13. Mohammadi M, Al-Fuqaha A, Sorour S, Guizani M. Deep learning for IoT big data and streaming analytics: a survey. IEEE Commun Surv Tutor. 2018;20(4):2923–60.

14. Li H, Wang H, Yin W, Li Y, Qian Y, Hu F. Development of remote monitoring system for henhouse based on IoT technology. Future Internet. 2015;7(3):329–41.

15. http://drrajivdesaimd.com/2016/07/19/internet-of-things-iot/

16. Montori F, Bedogni L, Bononi L. A collaborative Internet of Things architecture for smart cities and environmental monitoring. IEEE Internet Things J. 2018;5(2):592–605.

17. Sarkar C, et al. DIAT: a scalable distributed architecture for IoT. IEEE Internet Things J. 2014;2(3):230–9.

18. Kang K, Pang J, Xu LD, Ma L, Wang C. An interactive trust model for application market of the Internet of Things. IEEE Trans Ind Inf. 2014;10(2):1516–26.

19. Zeng X, et al. IOTSim: a simulator for analyzing IoT applications. J Syst Architect. 2017;72:93–107.

20. Kim M, Ahn H, Kim KP. Process-aware internet of things: a conceptual extension of the internet of things framework and architecture. KSII Trans Internet Inf Syst. 2016;10(8):4008–22.

21. Cuomo S, Somma VD, Sica F. An application of the one-factor Hull White model in an IoT fnancial scenario. Sustain Cities Soc. 2018;38:18–20.

22. Han SN, Crespi N. Semantic service provisioning for smart objects: integrating IoT applications into the web. Future Gener Comput Syst. 2017;76:180–97. 9

23. Alodib M. QoS-aware approach to monitor violations of SLAs in the IoT. J Innov Digit Ecosyst. 2016;3(2):197–207.

24. Adame T, Bel A, Bellalta B, Barcelo J, Oliver M. IEEE 802.11AH: the WiFi approach for M2M communications. IEEE Wirel Commun. 2014; 21(6):144–52.

25. Duttagupta S, Kumar M, Ranjan R, Nambiar M. Performance prediction of IoT application: an experimental analysis. In: Proc. 6th international conference on the internet of things, Stuttgart, Germany, 07–09 Nov 2016. p. 43–51.

26. Urbieta A, Gonzalez-Beltran A, Mokhtar SB, Hossain MA, Capra L. Adaptive and context-aware service composition for IoT-based smart cities. Future Gener Comput Syst. 2017;76:262–74.

27. Krishna GG, Krishna G, Bhalaji N. Analysis of routing protocol for low-power and lossy networks in IoT real time applications. Procedia Comput Sci. 2016; 87:270–4.

28. Jaber M, Imran MA, Tafazolli R, Tukmanov A. 5G Backhaul challenges and emerging research directions: a survey. IEEE Access. 2016; 4:1743–66.
29. Lianos, M. and Douglas, M. (2000) Dangerization and the end of deviance: the institutional environment. Brit J Criminol., 40, 261–278. http://dx.doi.org/10.1093/bjc/40.2.261.
30. Ashton, K.: That "Internet of Things" thing. RFID J. 2009 www.rfidjournal.com/article/print/4986.
31. R. Want, "An introduction to RFID technology," IEEEPervasive Comput. 2006; 5, No. 1, pp. 2533.doi:10.1109/MPRV.2006.2
32. L. Atzori, A. Iera and G. Morabito, "The Internet of Things: a survey," Comput Networks. 2010;54(15):pp. 2787–2805. doi:10.1016/j.comnet.2010.05.010.
33. M. Conti and S. Giordano, "Mobile ad hoc networking: Milestones, challenges, and new research directions," IEEE Commun. Mag., Vol. 52, No. 1, Jan. 2014, pp. 85–96.
34. M. J. McGlynn and S. A. Borbash, "Birthday protocols for low energy deployment and flexible neighbor discovery in adhoc wireless networks," in Proc. 2nd ACM Int. Symp. Mobile Ad Hoc Netw. Comput. (MobiHoc), Oct. 2001, pp. 137–145.
35. R. Roman and J. Lopez, "Integrating wireless sensor networks and the Internet: A security analysis," Internet Res. 2009;19(2):pp. 246–259.
36. Carnot Inst. (Jan. 2011). "Smart Networked Objects and Internet of Things," Carnot Institutes' Information Communication Technologies and Micro Nano Technologies Alliance, White Paper. [Online]. Available: http://www.internet-of-things-research.eu/pdf/ IoT_Clusterbook_March_2010.pdf, accessed 28 Nov 2011.
37. P. Bellavista, G. Cardone, A. Corradi, and L. Foschini, "Convergence of MANET and WSN in IoT urban scenarios," IEEE Sens J. 2013;13(10):pp. 3558–3567.
38. C. Perera, A. Zaslavsky, P. Christen, and D. Georgakopoulos, "Sensing as a service model for smart cities supported by Internet of Things," Trans Emerg Telecommun Technol. 2014;25(1):pp. 81–93.
39. E. Welbourne et al., "Building the Internet of Things using RFID: The RFID ecosystem experience," IEEE Internet Comput. 2009;13(3):pp. 48–55.
40. A. Zanella, N. Bui, A. Castellani, L. Vangelista and M. Zorzi, "Internet of Things for smart cities," IEEE Internet Things J. 2014;1(1): pp. 22–32.
41. A. M. Ortiz, D. Hussein, S. Park, S. N. Han, and N. Crespi, "The cluster between Internet of Things and social networks: review and research challenges," IEEE Internet Things J. 2014;1(3):pp. 206–215.
42. G. Quorum, F. Kawsar, D. Fitton, and V. Sundramoorthy, Smart objects as building blocks for the Internet of Things, IEEE Internet Comput. 2010;14(1):pp. 44–51. [Online]. Available: http://dx.doi.org/10.1109/MIC.2009.143.
43. Pacheco, J.; Satam, S.; Hariri, S.; Grijalva, C.; Berkenbrock, H. IoT security development framework for building trustworthy smart car services. In Proceedings of the IEEE Conference on Intelligence and Security Informatics, Tucson, AZ, USA, 28–30 September 2016; pp. 237–242.

44. Datta, S.K.; Bonnet, C. Describing things in the Internet of Things. In Proceedings of the IEEE International Conference on Consumer Electronics-Taiwan, Nantou, Taiwan, 27–29 May 2016; pp. 1–2.

45. Li, M.; Lin, H. Design and implementation of smart home control systems based on wireless sensor networks and power line communications. IEEE Trans. Ind. Electron. 2015;62:4430–4442.

46. Alam, M.R.; Reaz, M.B.I.; Ali, M.A.M. A review of smart homes—Past, present, and future. IEEE Trans Syst Man Cybern. 2012;42:1190–1203.

47. Fernandes, E.; Rahmati, A.; Jung, J.; Prakash, A. Security implications of permission models in smart-home application framework. IEEE Secur Priv. 2017;15:24–30.

48. Bugeja, J.; Jacobsson, A.; Davidsson, P. On privacy and security challenges in smart connected homes. In Proceedings of the European Intelligence and Security Informatics Conference, Uppsala, Sweden, 17–19 August 2016; pp. 172–175.

49. Parra, J.; Hossain, M.A.; Uribarren, A.; Jacob, E.; El Saddik, A. Flexible smart home architecture using device profile for web services: a peer-to-peer approach. Int J Smart Home 2009;3:39–56.

50. Garzon, S.R. Intelligent in-car-infotainment system: a prototypical implementation. In Proceedings of the 8th IEEE International Conference on Intelligent Environments, Guanajuato, Mexico, 26–29 June 2012; pp. 371–374.

51. Wollschlaeger, M.; Sauter, T.; Jasperneite, J. The future of industrial communication: automation networks in the era of the Internet of Things and Industry 4.0. IEEE Ind Electron Mag. 2017;11:17–27.

52. Bila, C.; Sivrikaya, F.; Khan, M.A.; Albayrak, S. Vehicles of the future: A survey of research on safety issues. IEEE Trans Intell Trans Syst. 2017;66: 2499–2512.

53. Mehrabani, M.; Bangalore, S.; Stern, B. Personalized speech recognition for Internet of Things. In Proceedings of the IEEE 2nd World Forum on Internet of Things, Milan, Italy, 14–16 December 2015; pp. 369–374.

54. Mumtaz, S.; Alsohaily, A.; Pang, Z.; Rayes, A.; Tsang, K.F.; Rodriguez, J. Massive internet of things for industrial applications: addressing wireless IIoT connectivity challenges and ecosystem fragmentation. IEEE Ind Electron Mag. 2017;11:28–33.

55. Internet of Things is a revolutionary approach for future technology enhancement: a review. Available online at https://link.springer.com/content/pdf/10.1186/s40537-019-0268-2.pdf. Accessed 20 Jun 2020.

56. Nusser, R.; Plez, R.M. Bluetooth-based wireless connectivity in an automotive environment. In Proceedings of the 52nd IEEE Vehicular Technology Conference, Boston, MA, USA, 24–28 September 2000; Volume 4, pp. 1935–1942.

57. 1. Kriufsky T.A, Kaplan L.M., "RFID inductive coupled device," Patent, United States of America, 1975, patent no. 3859624, pp. 1-3.

58. Beigal M, "RFID Device: A miniature size," Patent, United States of America, 1982, Patent Number 4333072, pp. 1-2.

59. Paul Zandbergen, "Short Range Wireless Communication: Bluetooth, Zigbee & infrared Transmission," Chapter 5, Lesson 7, Lesson Transcript, 2012, pp. 1-5. 4. Committee, "Zigbee Information," The revolution in connectivity, Viewed 2015, pp. 1–4. http://www.uei.com/applicationstechnology/connectivity

60. Jon Adams, "Busy as a Zigbee," IEEE Spectrum. 2006; pp. 1–3
61. 2015 International Conference on Green Computing and Internet of Things (ICGCIoT) 809
62. Stankovic, "Research directions for the Internet of Things," IEEE Internet Things J. 2014;1(1):pp. 3–9.
63. M. Smith et al., "RFID and the inclusive model for the IoT," CASAGRAS Partnership Rep., West Yorkshire, U.K., Final Rep., 2009, pp. 10–12.
64. Abidi, B.; Jilbab, A.; Haziti, M.E. Wireless sensor networks in biomedical: wireless body area networks. In Europe and MENA Cooperation Advances in Information and Communication Technologies; Springer: Berlin/Heidelberg, Germany, 2017; pp. 321–329.
65. Xu, Q.; Ren, P.; Song, H.; Du, Q. Security enhancement for IoT communications exposed to eavesdroppers with uncertain locations. IEEE Access. 2016;4: 2840–2853.
66. Lee JD, Yoon TS, Chung SH, Cha HS. Service-oriented security framework for remote medical services in the Internet of Things environment. Healthcare Inform Res. Oct. 2015;21(4):271–282.
67. Jaiswal S, Gupta D. Security requirements for Internet of Things (IoT). Proceedings; Singapore. Springer Singapore; 2017: 419–427.
68. Ahmed MU, Bjorkman M, Causevic A, Fotouhi H, Linden M. An overview on the Internet of Things for health monitoring systems. Proceeding of 2nd EAI International Conference on IoT Technologies for HealthCare. 2015 Oct 26–27; Rome, Italy. Springer; 2016: 429–436.
69. Zhang, W.; An, Z.; Luo, Z.; Li, W.; Zhang, Z.; Rao, Y.; Yeong, C.F.; Duan, F. Development of a voice-control smart home environment. In Proceedings of the IEEE International Conferenceon Robotics and Biomimetics, Qingdao, China, 3–7 December 2016; pp. 1697–1702.
70. Sarikaya, R.; Hinton, G.E.; Deoras, A. Application of deep belief networks for natural language understanding. IEEE/ACM Trans Audio Speech Lang Process. 2014;22:778–784.
71. BMW ConnectedDrive. Available online: http://intothefuture.eiu.com/how-will-connected-cars-workwith-smart-devices-in-your-home/. Accessed on 22 Apr 2017).
72. Liu, K.; Tolins, J.; Tree, J.E.F.; Neff, M.; Walker, M.A. Two techniques for assessing virtual agent personality. IEEE Trans Affect Comput. 2016;7:94–105.
73. Gartner. Gartner Says Smart Cities Will Use 1.6 Billion Connected Things in 2016. Available online: https://www.gartner.com/newsroom/id/3175418. Accessed: 20 Jun 2020.
74. Zanella, A.; Bui, N.; Castellani, A.; Vangelista, L.; Zorzi, M. Internet of things for smart cities. IEEE Internet Things J. 2014;1:22–32.
75. Samuel, S.S.I. A review of connectivity challenges in IoT-smart home. In Proceedings of the 2016 3rd MEC International Conference on Big Data and Smart City (ICBDSC), Muscat, Oman, 15–16 March 2016; pp. 1–4.
76. Ash, B.; Chandrasekaran, S. What the "Smart" in Smart Cities is All About. 2017. Available online: https://beyondstandards.ieee.org/smart-cities/smart-smart-cities/. Accessed 6 Mar 2018.
77. Shrouf, F.; Miragliotta, G. Energy management based on Internet of Things: practices and framework for adoption in production management. J Clean Prod. 2015;100:235–246.

78. Ejaz, W.; Naeem, M.; Shahid, A.; Anpalagan, A.; Jo, M. Efficient energy management for the Internet of Things in smart cities. IEEE Commun Mag. 2017;55:84–91.
79. L. Atzori, A. Iera, and G. Morabito, "SIoT: Giving a social structure to the Internet of Things," IEEE Commun Lett. 2011;15(11):pp. 1193–1195.
80. L. Atzori, A. Iera, G. Morabito, and M. Nitti, "The social Internet of Things (SIoT) – when social networks meet the Internet of Things: concept, architecture and network characterization," Comput Netw. 2012;56(16): pp. 3594–3608.
81. M. Al-Hader, A. Rodzi, A. R. Sharif, and N. Ahmad, "Smart city components architecture," in Proc. Int. Conf. Comput. Intell., Modelling Simulation, Sep. 2009, pp. 93–97.
82. N. Abbani, M. Jomaa, T. Tarhini, H. Artail, and W. El-Hajj, "Managing social networks in vehicular networks using trust rules," in Proc. IEEE Symp. Wireless Technol. Appl., Sep. 2011, pp. 168–173.
83. R. Fei, K. Yang, and X. Cheng, "A cooperative social and vehicular network and its dynamic bandwidth allocation algorithms," in Proc. IEEE Conf. Comput. Commun. Workshops, Apr. 2011, pp. 63–67.
84. S. Smaldone, L. Han, P. Shankar, and L. Iftode, "RoadSpeak: enabling voice chat on roadways using vehicular social networks," in Proc. 1st Workshop Social Netw. Syst., 2008, pp. 43–48.
85. S.Yousefi, E. Altman, R. El-Azouzi, and M. Fathy, "Analytical model for connectivity in vehicular ad hoc networks," IEEE Trans Veh Technol. 2008; 57(6):pp. 3341–3356.
86. Luo, P.; Zhang, M.; Ghassemlooy, Z.; Le Le Minh, H.; Tsai, H.-M.; Tang, X.; Png, L.C.; Han, D. Experimental demonstration of RGBLED-based optical camera communications. IEEE Photon J. 2015; 7:1–12.
87. Rachim, V.P.; Chung, W.-Y. Multilevel intensity-modulation for rolling shutter-based optical camera communication. IEEE Photon-Technol. Lett. 2018; 30:903–906.
88. Wang, W.-C.; Chow, C.-W.; Chen, C.-W.; Hsieh, H.-C.; Chen, Y.-T. Beacon jointed packet reconstruction scheme for mobile-phone based visible light communications using rolling shutter. IEEE Photon J. 2017;9:1–6.
89. Yang, Y.; Hao, J.; Luo, J. CeilingTalk: Lightweight indoor broadcast through LED-camera communication. IEEE Trans Mob Comput. 2017;16:3308–3319.
90. Hassan, N.B.; Ghassemlooy, Z.; Zvanovec, S.; Biagi, M.; Vegni, A.M.; Zhang, M.; Luo, P. Non-line-of-sight MIMO space-time division multiplexing visible light optical camera communications. J Light Technol. 2019; 37:2409–2417.
91. Hasan, M.K.; Chowdhury, M.Z.; Shahjalal; Nguyen, V.T.; Jang, Y.M. Performance analysis and improvement of optical camera communication. Appl Sci. 2018; 8:2527.
92. Burgess, P. Adafruit Neopixel Uberguide, WS2812B Datasheet; Philips, The Netherlands, 2019. Available online: https://cdn.learn.adafruit.com/downloads/pdf/adafruit-neopixel-uberguide.pdf. Accessed 13 Jun 2020.
93. Fujihashi, T.; Koike-Akino, T.; Orlik, P.; Watanabe, T. High-throughput visual MIMO systems for screen-camera communications. IEEE Trans Mob Comput. 2020.

94. Jerkovits, T.; Liva, G.; IAmat, A.G. Improving the decoding threshold of tail-biting spatially coupled LDPC codes by energy shaping. IEEE Commun. Lett. 2018; 22:660–663.

95. Fang, Y.; Chen, P.; Cai, G.; Lau, F.C.M.; Liew, S.-C.; Han, G. Outage-limit-approaching channel coding for future wireless communications: root-protograph low-density parity-check codes. IEEE Veh Technol Mag. 2019; 14:85–93.

96. Elshabrawy, T.; Robert, J. Interleaved chirp spreading LoRa-based modulation. IEEE Internet Things J. 2019;6:3855–3863.

97. Shahjalal; Hasan, M.K.; Chowdhury, M.Z.; Jang, Y.M. Smartphone camera-based optical wireless communication system: requirements and implementation challenges. Electronics. 2019;8:913.

98. Mumtaz, S.; Alsohaily, A.; Pang, Z.; Rayes, A.; Tsang, K.F.; Rodriguez, J. Massive Internet of Things for industrial applications: addressing wireless IIoT connectivity challenges and ecosystem fragmentation. IEEE Ind Electron Mag. 2017.

99. 7 ways AI is impacting the automotive industry. Available at https://techstory.in/7-ways-ai-is-impacting-the-automotive-industry/. Accessed 19 Jun 2020.

100. Paul, G.; Thoma, D.; Irvine, J. Privacy implications of smartphone based connected vehicle communications. In Proceedings of the 82nd IEEE Vehicular Technology Conference, Boston, MA, USA, 6–9 September 2015; pp.1–3.

101. Wang, H.; Saboune, J.; Saddik, A.E. Control your smart home with an autonomously mobile smartphone. In Proceedings of the IEEE International Conference on Multimedia and Expo Workshops, San Jose, CA, USA, 15–19 July 2013; pp. 1–6.

102. Li, Y.; Zhuang, Y.; Lan, H.; Zhang, P.; Niu, X.; El-Sheimy, N. Self-contained indoor pedestrian navigation using smartphone sensors and magnetic features. IEEE Sens. J. 2016;16:7173–7182.

103. Ellison, G.; Lacy, J.; Maher, D.P.; Nagao, Y.; Poonegar, A.D.; Shamoon, T.G. The car as an Internet-enabled device, or how to make trusted networked cars. In Proceedings of the IEEE International Electric Vehicle Conference, Greenville, SC, USA, 4–8 March 2012; pp. 1–8.

104. SAE. Convergence 2014 survey: automated, connected, and electric vehicle systems. 2014. Available online: http://graham.umich.edu/media/files/LC-IA-ACE-Roadmap-Expert-Forecast-Underwood.pdf (accessed on 22 April 2017).

105. Telefnica. Connected car industry report, 2013. Available online: https://iot.telefonica.com/system/files_force/telefonica_digital_connected_car_report_english_36.pdf?download=1. Accessed 22Jun 2020.

106. He, W.; Yan, G.; Wu, L.D. Developing vehicular data cloud services in the IoT environment. IEEE Trans. Ind. Inform. 2014;10:1587–1595.

107. Gerla, M. Vehicular cloud computing. In Proceedings of the 11th IEEE Annual Mediterranean Ad Hoc Networking Workshop, Ayia Napa, Cyprus, 19–22 June 2012; pp. 152–155.

108. Yan, G.; Wen, D.; Olariu, S.; Weigle, M.C. Security challenges in vehicular cloud computing. IEEE Trans Intell Trans Syst. 2013;14:284–294.

109. Gubbi, J.; Buyya, R.; Marusic, S.; Palaniswami, M. Internet of things (IoT): a vision, architectural elements, and future directions. Future Gener Comput Syst. 2013;29:1645–1660.

110. Datta, S.K.; Costa, R.P.F.D.; Härri, J.; Bonnet, C. Integrating connected vehi-cles in Internet of Things ecosystems: challenges and solutions. In Proceedings of the IEEE 17th International Symposium on A World of Wireless, Mobile and Multimedia Networks, Coimbra, Portugal, 21–24 June 2016; pp. 1–6.
111. Salman, O.; Elhajj, I.; Kayssi, A.; Chehab, A. Edge computing enabling the Internet of Things. In Proceedings of the 2015 IEEE 2nd World Forum on Internet of Things (WF-IoT), Milan, Italy, 14–16 December 2015; pp. 603–608.
112. Wang, Y.; He, S.; Mohedan, Z.; Zhu, Y.; Jiang, L.; Li, Z. Design and evalua-tion of a steering wheel-mount speech interface for drivers' mobile use in car. In Proceedings of the IEEE Conference on Intelligent Transportation Systems (ITSC), Qingdao, China, 8–11 October 2014; pp. 673–678.

Chapter 2

Accident warning and collision avoidance systems

Pooja Jha

Amity University Jharkhand, Ranchi, India

CONTENTS

2.1 INTRODUCTION

Road accidents are an unpredictable facet of our lives. According to one study, by 2030, road accidents are predicted to be the fifth leading cause for human fatalities [1]. According to another study, injuries from road accidents are the twelfth leading cause for human disabilities [2]. Road accidents are the chief cause of injuries and death in teens [3], and they are the second leading cause of deaths in countries such as Europe [4]. Based on these statistics, we can conclude that is impossible to avoid the road

DOI: 10.1201/9781315110905-2

accidents; however, we can minimize them using the latest developments in the field of Intelligent Transport System (ITS), like autonomous vehicles (AVs).

AVs can prove to be very helpful in avoiding road accidents. There are many reasons for this, such as AVs do not drink or become distracted like human drivers, and, therefore, they have fewer chances of accidents when compared with human-driven vehicles [5]. AVs can reduce collisions due to their enhanced insight (for example, they have no blind spots), appropriate responses, and speedy execution of actuators such as steering components, brakes, and gas pedals [1]. In addition, the number of collisions can be decreased by introducing inter-AVs and Roadside Units (RSUs) based on communication capabilities [6]. Google AVs had fewer collisions when compared with human-driven vehicles between the years of 2009 to 2015 [7]. Though, in most investigations, collision avoidance (CA) has concentrated on, respectively, the rear end, the front end, and lateral collisions in areas that are less congested than the open road with long intervehicular distances.

Although it had been a century since the invention of automobiles, transportation by land remains the most hazardous means to travel and transport goods. According to the Bureau of Transportation Statistics, in 2012, the total number of vehicles in the United States increased to 254 million [8]. Improvements in safety measures are made constantly; consequently, the rate of deaths caused by car accidents has been reduced over time. Many improvements have been made to cars since the first vehicle was built. Today, cars have more powerful and efficient engines, better dynamics, more controllability, and greater stability on the road. Conversely, the total number of car-accident victims still remains high. Every year, approximately 33,000 people die on the road [9]. Although cars are safer and easier to handle, they require greater concentration from the driver. Heavy traffic, high speeds, and fast maneuvers require drivers to fully concentrate and react swiftly. Humans have a natural limit that does not permit them to always react as quickly as needed and take vital measures when they are required. Current development shows aiding drivers broaden their reactions and driving capabilities [10]. Advanced sensors that aid drivers' visibility in the dark, detectors, and fast control logic have already made a mark in this field. Another improvement is the driving advising system, which monitors the activity of the driver and environment, and takes control of the car when it is needed [11, 12]. One optimistic prediction is that the use of fully AVs may minimize the total number of accidents by 90 percent and lead to enormous economic advances in this industry [13, 14]. This encourages leaders in the industry to develop and introduce an AV.

However, there are many challenges to tackle given developing an AV is a complex task. One challenge is localization and mapping. It is not only difficult to determine the exact position of a vehicle on the road, but also any road work changes the existing map's route and can make it difficult or impossible to follow. Another challenge is sensing. Sensing is a broad topic that relates to computer vision and object recognition from the sensor's data. Yet another challenge relates to controlling the AV, especially with interaction from human drivers. The transition from now to the days when all vehicles on the road are autonomous may take decades. Until then, both human drivers and AVs will be present on the road and must be able to coexist with each other.

2.2 FEATURES AVAILABLE IN AUTONOMOUS VEHICLES

This section discusses the various features that distinguish an AV from a non-AV vehicle. Let us begin with the first feature, which is a sensor.

2.2.1 Sensors

AVs use sensors to identify, categorize, and/or localize pedestrians in the street environment. Some types of sensors reliably detect that something significant is on the road but may have difficulty determining what the object is. Other sensors classify objects into separate categories, such as pedestrians and bicyclists, but only after they have been detected by other means or under ideal conditions. Classification is useful in predicting the likely movement of objects in the streetscape, because objects such as pedestrians, bicyclists, and garbage cans behave differently. Therefore, the capacity of a sensor to identify objects precisely and follow their positions in three-dimensional space is crucial for the driving system to ascertain whether it needs to brake or change lanes.

Next, the camera sensors, namely stereo vision and object recognition, are discussed.

2.2.1.1 Camera sensors: stereo vision

A duo of front-end cameras can offer an autonomous system three-dimensional depth perception of the situation ahead. Video-processing algorithms ascertain in what way characteristics such as edges and consistencies relate between the two camera illustrations; the parallax between these characteristics is subsequently used to assess the scope for every pixel spot in the scene. Pixel-level range resolution delivers an accurate reading of the angle of an object, although the accuracy of the range estimation declines with distance. Stereo vision systems (see Figure 2.1) can be very efficient in identifying and concentrating on visible, human-sized objects in the streetscape irrespective of the systems' capability to categorize them. If a human-sized object appears in the center of the highway, an autonomous system can be programmed to prevent driving into it irrespective

Figure 2.1 Subaru's EyeSight automatic emergency braking system uses stereo cameras mounted above the rearview mirror [38].

of categorization. The main constraint of stereo vision systems is the changing degree of object reflectivity at a distance under different lighting, weather, and reflectance circumstances [38].

2.2.1.2 Camera sensors: object recognition

Several camera-equipped systems utilize pattern-recognition algorithms to categorize objects as road signs, traffic barrels, cars, trucks, motorcycles, bicyclists, and pedestrians. The fine pixel resolution of an illustration enables the system to create accurate angular localization of recognizable objects. The range might be projected accurately from stereo disparity if two cameras are used; otherwise a monocular system must approximate the range roughly from projected object sizes or displacement from the horizon. The latest innovations in machine-learning techniques have made object-categorization systems much more efficient than they were just a few decades ago.

If the vision algorithm is extremely tolerant, it can incorrectly categorize contextual chaos, such as individuals, which could cause needless and theoretically risky emergency decelerating [38]. On the contrary, if the computer-vision algorithm fails to identify an actual individual or vehicle in a location, the driving system is sightless unless a superfluous sensing method can distinguish the possibility for collision. Today, computer-vision algorithms are still restricted enough that AVs cannot operate on video unaided, particularly with only a monocular camera.

2.2.2 Automotive radar

Radar systems actively sense the range of objects by transmitting a continuously varying radio signal and measuring reflections from the environment. A computer compares the incoming and outgoing signals to estimate both the distance and relative approach velocity of objects with high accuracy.

Determining the angle to objects is more challenging. Instead of using a rotating antenna like an airport radar, automotive radar systems scan across the sensing area by using an array of electronically steered antennas. *Beamforming* is a technique used to aim the effective direction of a radar beam by changing the delay between different transmitting antennas so that the signals add constructively in a specified direction. Multiple receiving antennas allow further angular refinement. However, the resulting radar beam is still wide enough that the resulting angular resolution is limited. Most radar systems are insufficient for making decisions that require precise lateral localization, such as whether there is enough room to pass a bicyclist or pedestrian ahead on the roadway edge without changing lanes.

Pedestrian detection and classification with radar can be enhanced through the use of micro-Doppler signal analysis. The movement of a pedestrian's arms and legs while walking provides a unique time-varying Doppler signature that can be detected in the radar signal by a pattern-recognition algorithm. This signature allows the radar system to distinguish pedestrians from other objects and to start tracking their movements sooner and among clutter, such as when a pedestrian steps out from between parked cars. Bicyclists present more of a challenge to distinguish from clutter because they move their body less, especially when coasting [38].

2.2.3 Sensor fusion

The accuracy of object detection, classification, and localization improves by combining data from complementary sensors. Cameras provide good angular resolution, while radar provides good range precision; vision algorithms can classify visible objects, but radar can detect obstacles in darkness and glare. Volvo has used this combination for pedestrian and bicyclist CA since 2013. Ford Motor Company's marketing campaign for the 2018 Mustang has highlighted Ford's use of radar/video sensor fusion to enhance the vehicle's automatic emergency braking performance, particularly for pedestrians and at night [38].

2.2.4 Lidar

Lidar (Light Detecting and Ranging) sensing uses time-of-flight measurements from reflected laser beams to measure objects and roadway features around the vehicle. The Lidar offers the angular precision of an optical system with the range precision of radar. Lidar sensors have a reputation for being large, but recent disruptive innovations in Lidar technology are poised to make it practical for widespread automotive use (Figure 2.2).

Lidar uses rapidly scanning infrared laser beams to generate a 3D point cloud of measurements with sufficient resolution to detect, classify, and localize cars, pedestrians, bicyclists, and even small animals around the vehicle [38]. Due to the three-dimensional nature of the data, Lidar systems can detect and

Figure 2.2 Compact solid-state automotive Lidar sensors are relatively inconspicuous and approaching the affordability of camera and radar sensors [38].

track a human-sized or larger object in the streetscape, even before classifying it as a motorcyclist, bicyclist, or pedestrian. As the distance between the vehicle and object decreases, the increased number of points on the object makes classification based on shape possible, although vision sensors may also play a role in classification. Lidar sensors are unaffected by darkness and glare and suffer relatively minor degradation in rain, snow, and fog.

The next section discusses the CA system in AVs.

2.3 COLLISION AVOIDANCE ASSISTANCE (CAA)

Collision-related safety systems [19] can be categorized as passive and active safety systems based on their deployment. Such systems have traditional seatbelt and airbag systems that reduce injury to the driver and passenger if a collision occurs. They also have rescue systems, such as the system that involuntarily alerts rescue centers about the scene of a crash [15, 16]. CA systems are predominately centered on pre-crash functions [17, 18] that alert and support the driver, and control the vehicle in hazardous conditions. Such systems are essential in automatic driving systems. Although automatic CA systems cannot handle complicated traffic situations with precision yet, some capabilities are present in the sensing technologies. In addition, automatic driving technologies are still in their infancy, and a considerable improvements need to be made prior to their wide-spread use.

Collision avoidance assistance (CAA) systems aid drivers in averting barriers at high-speeds by using the traditional driver-in-loop driving systems. In such systems, the driver maintains the uppermost level of command of the vehicle and dominates the system. Hence, false system triggers are avoided, thereby improving the system's safety and reliability. There are many CAA systems used today, such as the collision warning (CW), lane departure warning (LDW), lane keeping assist (LKA), and automatic emergency brake (AEB) systems [20] The CW and LDW systems alert the driver to potential collision dangers, but they do not recommend any action for evading such collisions or controlling the vehicle. The LKA system acts as soon as the system identifies that the vehicle is about to diverge from a traffic lane due to the driver's lack of attention or exhaustion, thereby employing a slight amount of counter-steering force. The AEB system uses the brakes spontaneously in reaction to the recognition of a probable collision and prevents the longitudinal potential collision.

The number of vehicles in the world has increased exponentially in the twenty-first century, leading to an increase in the number of vehicles on the road and number of drivers [21]. Therefore, issues like how to reduce the number of traffic accident deaths and economic losses have become more prevalent [22].

Research has been done on the collision warning algorithm. Current collision warning algorithms are primarily split into two types, namely, the Safety Time Algorithm and the Safety Distance Algorithm [23]. These algorithms assess the collision time using the safety time limit to determine the condition of safety. The algorithm primarily uses Time to Collision (TTC) [24]. The Safety Distance model applies to the shortest distance between the vehicle and the barrier; in other words, this is the distance the vehicle requires in order to avoid colliding with the obstacle [25].

Recent research on AVs focuses on how to achieve safety and ease for the drivers [26]. Even though a completely AV, without a driver, has not yet developed, a semi-autonomous vehicle has been already mass-produced, with a partially automatic system for braking and steering [27]. The techniques and knowledge used to build full AVs have kept in mind how to safeguard the drivers from any fatal accidents.. Semi-autonomous automotive products, such as Advanced Driver Assist System (ADAS), which includes an autonomous emergency brake system, adaptive cruise control, lane keeping assist, active blind spot detection, and autonomous parking assist are available. Also, an oncoming vehicle CA system that uses automated steering and a front camera for oncoming vehicles has been developed recently, as shown in Figure 2.3.

The oncoming vehicle CA system is helpful because it automatically avoids traffic accidents with oncoming vehicles. It controls the electric power steering of the host vehicle without applying the brake when a front camera detects oncoming vehicles with a risk of collision. It helps to minimize head-on collision accidents that can lead to deadly damages to drivers. Hence, various companies from different parts of world are competing to advance the system to protect drivers and passengers from deadly crashes [28, 29]. These systems are high performance to address robust conditions, such as high speeds. High performance sensors are required to identify oncoming vehicles and control the electric power steering. Specifically, the

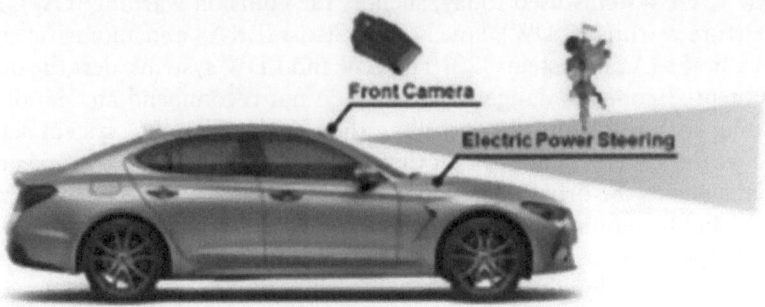

Figure 2.3 Oncoming vehicle CA system.

system should avoid collision with vehicles at high speeds because the relative speed is greater with oncoming vehicles.

2.4 ALGORITHMS USED IN ACCIDENT AVOIDANCE—A REVIEW

This section of the chapter discusses some of the accident warning and CA systems in brief. It begins with CA using a vehicular cyber-physical system (VCPS) [1].

2.4.1 Collision avoidance using VCPS

This section explains a road-based taxonomy briefly, as shown in Figure 2.4.

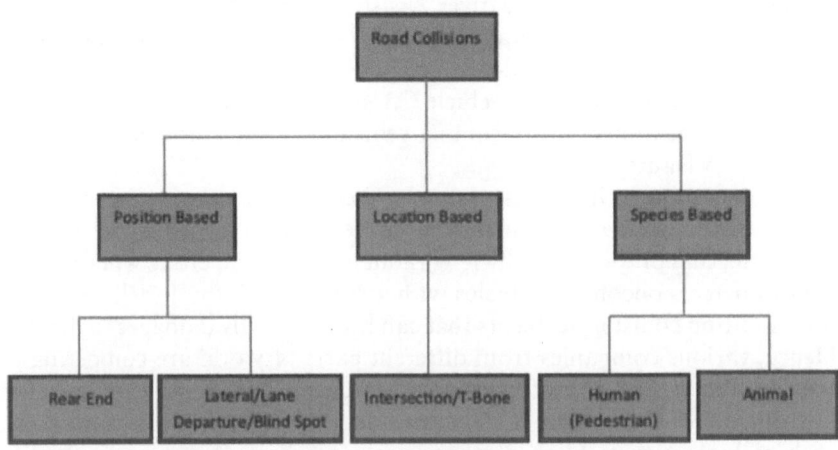

Figure 2.4 Road collision taxonomy.

2.4.1.1 Position based

Position-based collisions can be divided into two subtypes: rear end and lateral/lane departure.

2.4.1.1.1 Rear end

Definition: A rear-end collision [39] is a transportation accident from where an agent attacks the back of another agent or vehicle. Figure 2.5 below shows the condition for rear end collision.

2.4.1.1.2 Lateral/lane departure/blind spot

Definition: In a lateral collision, two vehicles traveling in parallel directions converge, colliding with each other side by side (Figure 2.6) [40].

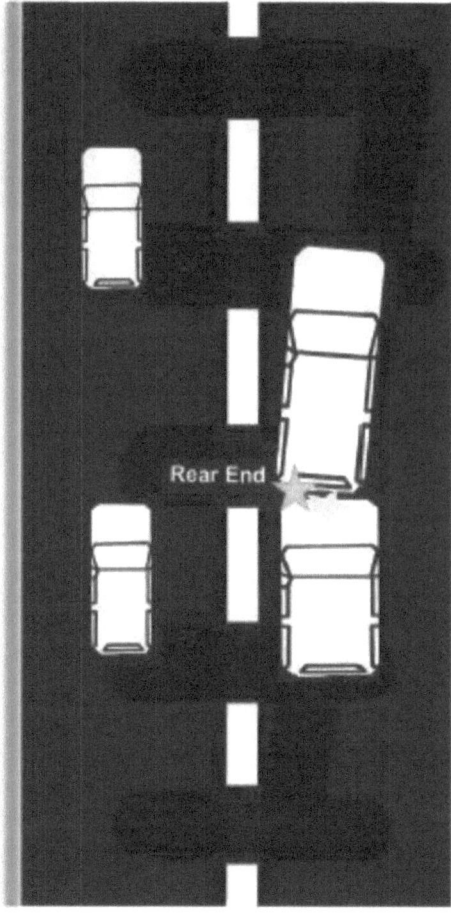

Figure 2.5 Rear-end collision scenario.

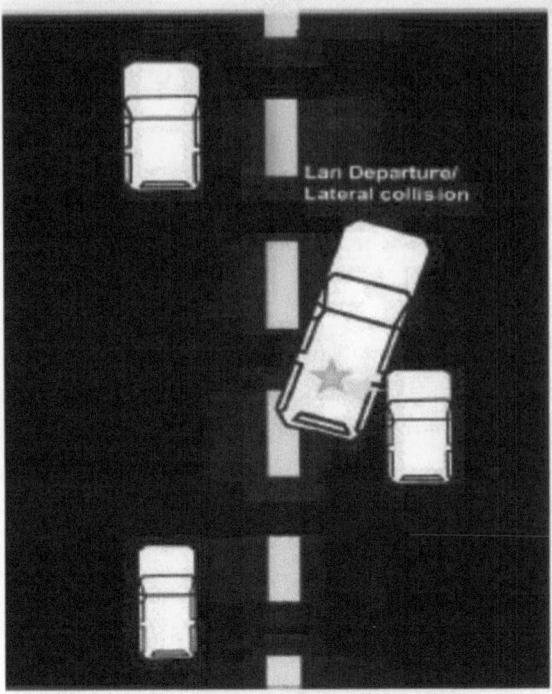

Figure 2.6 Lateral/lane departure/blind spot collision scenario.

2.4.1.2 Location based

In this section, location-based road collisions are described. Intersection/T-bone collisions fall under location-based collisions.

2.4.1.2.1 Intersection/T-bone

Definition: In a T-bone collision, one vehicle collides with the side of another vehicle in a perpendicular fashion due to the violation of a red light or stop signals at an intersection [41] (see Figure 2.7).

The next section discusses the benefits and limitations of AVs.

2.5 BENEFITS OF V2V

1. **Avoids crashes:** Around 32,000 people are killed in car collisions worldwide every year. This number is continuously rising with time. Safety has become the most important issue. Despite efforts to boost consciousness and instruct people on safe driving, the major reason of car accidents remains human error. Vehicle-to-Vehicle (V2V)

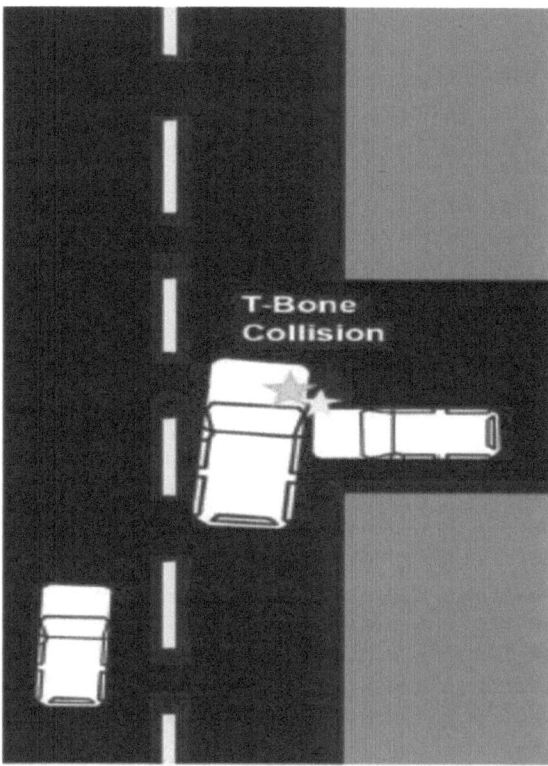

Figure 2.7 T-bone collision scenario.

communication technology can help reduce up to 70 percent to 80 percent of vehicle crashes involving human error [30].

2. **Enhances traffic management and minimizes congestion:** Law enforcement can use V2V transmission to monitor and control traffic by using real-time data streaming from vehicles to reduce congestion. V2V communication can assist in re-routing traffic, tracking vehicle locations, adapting traffic light schedules, and addressing speed limits. Drivers using V2V communication can prevent traffic jams and maintain a safe distance from other cars.

3. **Enhances fuel efficiency:** V2V communication facilitates fleets of trucks to drive in close formation. The trucks in the platoon remain in constant formation and adapt their velocity and position based on a constant stream of communication.

4. **Enhances directions:** Once V2V communication technology will be fully adopted commercially, every vehicle on the road will benefit from better navigation. Open-channel communication between all vehicles will provide precise locations, speed, and positioning information that will help each vehicle optimize routes in real time.

2.6 LIMITATIONS OF V2V

Numerous factors [30] deter the implementation of V2V communication. Commercially incorporating the technology poses questions for global, public, and private sectors, from security concerns to protocol guidelines.

1. **Security consequences:** The door lock, the wheel sending the driver on a spin, the car engine accelerating and crossing the speed limit are some of the security concerns in V2V communications. Also, the chance that the vehicles are vulnerable to cyberattacks is high. V2V communication systems will involve thorough security procedures to be fully integrated.
2. **Privacy issues:** The V2V network gathers and stores personal data about drivers. Because regulations have not existed until now, the government and private companies can track vehicles and monitor driving habits. Anyone with access will be able to track and collect data about cars with V2V communication. If the data is hacked, it can lead to identity theft and other security concerns.
3. **Obligation anxieties:** Since V2V technologies are still new and there are not clear laws and regulations, incidents involving V2V vehicles may result in liability concerns. The question about what instructions will be given by the V2V communication system when it leads to an accident remains; that is, the driver can argue he was only following the system's instructions when he crashed into the back of a car.
4. **Potentially distracting drivers:** At this time, V2V communication systems need human intervention to work. The driver needs to perform tasks like texting or talking on the phone to operate the V2V communication system. The communication process is still in the works, and it will need to be less distracting to the driver or it may end up being a new cause of traffic accidents.
5. **Expensive:** The cost of installing V2V communication systems in the vehicles depends on the system's complexity and vehicle model.

Next, the chapter discusses some cases where operating an AV proved to be its failure.

2.7 CASES OF FAILURE OF AVs

This section lists some examples of failures in AVs. A case with Uber is discussed first [31].

2.7.1 The autonomous Uber

In March of 2018, pedestrian Elaine Herzberg was killed by one of Uber's self-driving cars as she crossed a multi-lane road in Tempe, Arizona. The investigation included data from the car as well as dashcam video.

While it was determined that the automatic braking system was turned off to avoid erratic driving conditions, the human driver, Rafaela Vazquez, was found at fault for the death. Dashcam video showed Vasquez repeatedly looking down at her lap in the final minutes before the crash, including the five seconds before impact. If the driver had been paying attention, the car's data showed that the accident most likely could have been avoided.

2.7.2 Tesla on autopilot

In March 2019, Jeremy Banner's Tesla Model 3 collided with a tractor-trailer that was crossing his path in a Florida highway. An investigation is currently underway, as his family is suing Tesla for wrongful death. A preliminary report from the National Transportation Safety Board revealed Banner turned on autopilot just before the crash, and the vehicle "did not detect the driver's hands on the steering wheel." This incident was like a 2016 accident, which killed 40-year-old Joshua Brown, who was using autopilot when a tractor-trailer crossed his path. Tesla said during that investigation that its camera system failed to recognize the white broadside of the truck against the bright sky.

2.7.3 Tesla and the 2-year-old driver

In December 2018, Mallory Harcourt of Santa Barbara, claims that, while unloading groceries from her Tesla Model X, parked in the driveway, her two-year-old son jumped in the driver's seat, and the car unexpectedly lurched forward, ultimately pinning her to the garage wall. Harcourt, who was pregnant, suffered a broken leg and pelvis, and went into labor, which led to the premature birth of her daughter.

2.8 ALGORITHMS/MECHANISMS FOR ACCIDENT AVOIDANCE

The next section covers some of the algorithms that have been developed recently for avoiding accidents that may occur with AVs. The pedestrian CA system is discussed [32]. Before going into the logic of algorithms, however, let us discuss the schematic diagram of the pedestrian CA algorithm (see Figure 2.8).

Figure 2.8 Schematic diagram of pedestrian CA algorithm.

The mechanism can be summarized as follows [32]:

First, the projected pedestrian situation is built on the detected place by a Lidar and a constant- velocity pedestrian motion model. Next, the situational risk assessment algorithm, which depends on potential fields, establishes the desired deceleration to avoid the accident. Finally, the velocity of the vehicle is regulated by applying the active control of the in-wheel-motor torque. The projected pedestrian position is computed. This is the pedestrian position when the vehicle goes beyond the conflict point between the vehicle's moving axis and the pedestrian's moving axis on the crosswalk in the intersection.

Potential field is used as the basic algorithm to decide the preferred acceleration to prevent the accident [32]. To deal with the fact that barriers cannot be sensed in advance, the danger preventative mechanism is presented in the autonomous braking control system design. The predicted pedestrian position was built on the existing edge point of a pedestrian crossing the closure area by sensors such as Lidar, and a constant-velocity pedestrian motion model. If some closure is predictable, the position of the closure will be used to envisage the appearance of the pedestrian. Next, the situational risk assessment algorithm based on the potential field method determines the desired deceleration to slow down the vehicle and avoid the collision with the pedestrian.

Another algorithm discussed is based on a sensing infrastructure detecting road geometry and a two-layer accident-avoidance framework consisting of a threat assessment and an intervention layer [33]. An innovative active safety function for inhibiting the loss of vehicular control is implemented using the considered accident-avoidance framework. Under this framework, the accident-avoidance framework in the system and a novel active safety function was proposed that was based on road-preview information [34]. It prevents the vehicle from operating in conditions where assistance from an Electronic Stability Control (ESC) system is normally needed. A novel decision-making algorithm that can switch between three different intervention strategies was derived. The safety function discussed here relies less on the driver's steering action than ESC. In addition, the capability of intervening before the vehicle was already deviating from its nominal behavior increases the possibility to keep the vehicle on the road in critical situations. Figure 2.9 below represents the suggested accident prevention architecture.

CA systems, on the other hand, fall in the category of safety applications and automatically take control of the vehicle's motion, if needed [35]. The transition criteria, from manual to automated or semiautomated mode, are thus inherently different in safety and assistance/comfort applications.

There are two layers, the threat-assessment layer and the intervention layer, which utilize information from the environment information module that provides estimates of factors such as the road geometry, surface adhesion, and/or distances to other objects. A state estimation and parameter identification module provides estimates of the vehicle state and, possibly, driver and vehicle model parameters. In the intervention layer, an

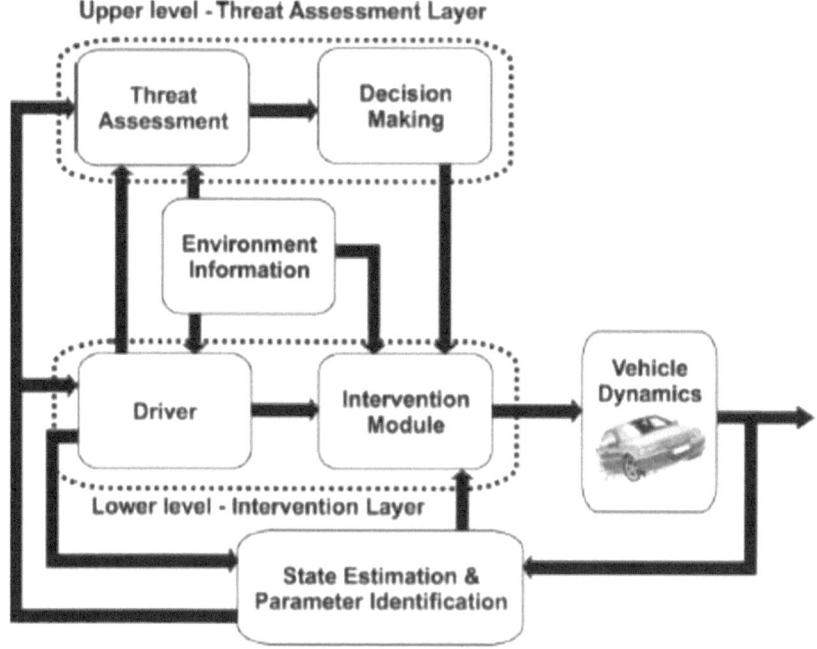

Figure 2.9 Overview of the suggested accident prevention architecture.

intervention module implements a set of intervention strategies. Available intervention strategies might include the activation of an ESC controller, a steering controller for keeping the vehicle in the lane, or a completely automated driving system where steering, braking, and acceleration are automatically controlled. The switching between intervention strategies is decided in the threat-assessment layer, where a distinction between the threat-assessment and decision-making modules is made. The threat-assessment module evaluates the risk of an accident for the different intervention strategies, whereas the decision-making module decides which intervention should be activated [33].

Another algorithm/mechanism is able to avoid vehicles or obstacles by proposing a smooth, local modified trajectory of a global path [36].

Yet another resource presents a local path modification of a global reference trajectory [37]. It proposes a safe trajectory by avoiding static and dynamic obstacles thanks to a lane change maneuver. The initial trajectory can be provided from a global path planner using a digital map database or a recorded map. The architecture to control the AV is generally composed of different steps (shown in Figure 2.10), which are perception (including localization), trajectory generation, and control.

The proposed algorithm is based on the two previously mentioned principles by uses a configurable sigmoid function to provide an efficient and

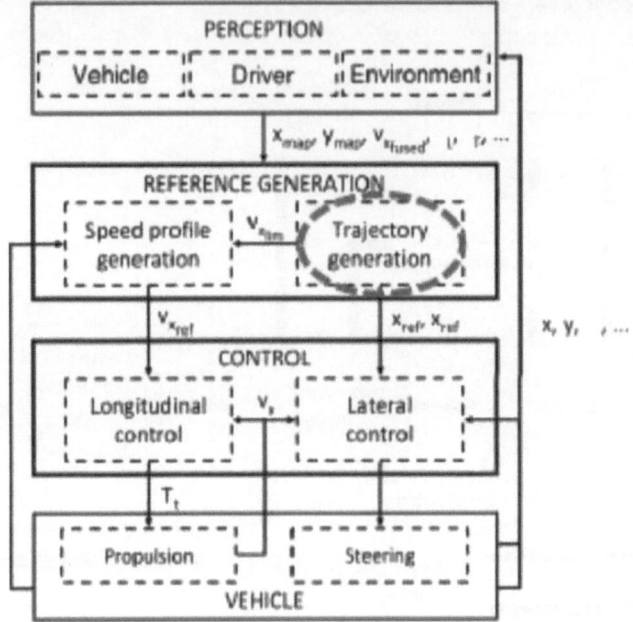

Figure 2.10 Architecture of the control of the vehicle proposed by [37].

feasible path to the vehicle in each horizon. The path takes into consideration the position and the motion of the perceived obstacles or the leading vehicle. The algorithm is constituted of different parts that are able to react to any detected obstacle on the road. The initial reference path of the vehicle can be obtained from a digital map database or a recorded map. The inputs of the algorithm are the vehicle position and the obstacle positions, which represent the perception and localization parts (shown in Figure 2.11). The local path is then calculated depending on the desired parameter of smoothness and the two parameters of safety distance to be observed during the avoidance maneuver. This parametric smooth variation is necessary to maintain a level of safety and comfort for the passenger of the vehicle and to respect the vehicle dynamics when doing a maneuver. The safety distance when avoiding obstacles can be configured by taking into consideration the vehicle velocity. The waypoint calculated represents the position to reach when avoiding the obstacle/vehicle based on the generated avoidance path.

2.9 CONCLUSIONS

Road accidents are caused due to numerous reasons. The goal of the chapter is to provide an extensive overview of the work that has been done related to CA solutions. *Artificial intelligence* refers to machines that are

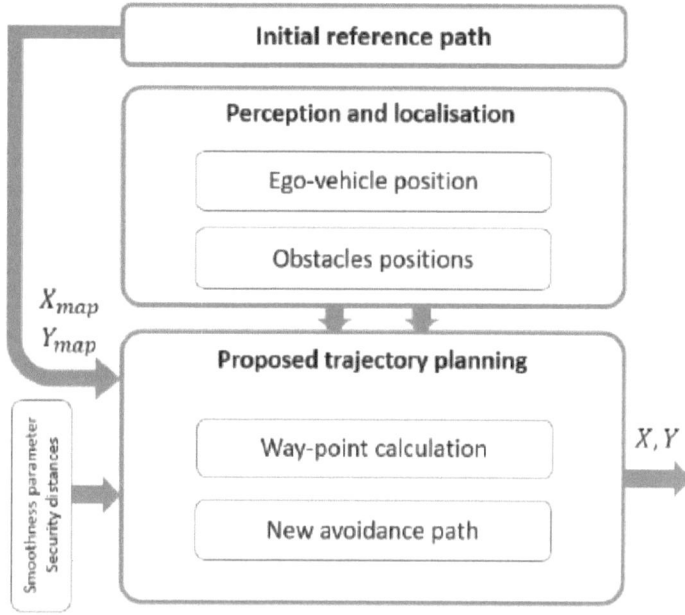

Figure 2.11 Path-planning strategy.

built to act like human beings, by studying human beings. AVs are in town and no one can negate their importance. However, building collision-free AVs is a challenging task. To enhance safety in intersections, the chapter has highlights the work done on the right-turn driving maneuver, which drivers need to negotiate pedestrians near, or in, crosswalks. The chapter highlights how engineers have employed potential field as the basic algorithm to determine the desired acceleration to avoid a collision. To deal with situations where obstacles cannot be detected beforehand, a hazard anticipatory mechanism is introduced in the autonomous braking control system design. The autonomous CA algorithm was created to prevent a collision with pedestrians near the crosswalk. Prototypes for various system components are currently being manufactured and tested against presupposed functionalities. Lab and in-field experiments are being conducted to evaluate system performance in simulation and in real life. The best way to reduce uncertainty while maintaining the same flow is to reduce localization error, which unfortunately is not always feasible. Other ways to reduce uncertainty include reducing velocity or increasing intervehicular distances, or both, but these usually impact the flow capacity. Collision avoidance with a vehicle, a pedestrian, or any obstacle and the generation of a feasible continuous curvature trajectory represent the major problems researchers face as they work to provide a safe path-planning solution.

REFERENCES

1. Riaz F, Niazi MA. Road collisions avoidance using vehicular cyber-physical systems: a taxonomy and review. Complex Adaptive Systems Modeling. 2016;4(1): 15.
2. Gopinath B, Harris IA, Nicholas M, Casey P, Blyth F, Maher CG, et al. A comparison of health outcomes in older versus younger adults following a road traffic crash injury: a cohort study. PLoS one. 2015;10(4): e0122732.
3. Taubman–Ben-Ari O, Kaplan S, Lotan T, Prato CG. Parents' and peers' contribution to risky driving of male teen drivers. Accident Analysis & Prevention, 2015;78: 81–86.
4. Copsey N, Drupsteen L, Kampen Jv, Kuijt-Evers L, Schmitz-Felten E, Verjans M. A review of accidents and injuries to road transport drivers. EU-OSHA, 2010.
5. Macy ML, Carter PM, Bingham CR, Cunningham RM, Freed GL. Potential distractions and unsafe driving behaviors among drivers of 1- to 12-year-old children. Academic pediatrics, 2014;14(3): 279–86.
6. Baskar LD, De Schutter B, Hellendoorn J, Papp Z. Traffic control and intelligent vehicle highway systems: a survey. IET Intelligent Transport Systems, 2011;5(1): 38–52.
7. Blanco M, Atwood J, Russell S, Trimble T, McClafferty J, Perez M. Automated Vehicle Crash Rate Comparison Using Naturalistic Data. Virginia Tech Transportation Institute, 2016.
8. Bureau of Transportation Statistics. Table 1-11: Number of U.S. Aircraft, Vehicles, Vessels, and Other Conveyances. Bureau of Transportation Statistics, May 2013.
9. Blincoe LJ, Miller TR, Zaloshnja, E, and Lawrence BA, The economic and societal impact of motor vehicle crashes. U.S. Department of Transportation, 2015.
10. Special Investigation Report. The use of forward collision avoidance systems to prevent and mitigate rear-end crashes, National Transportation Safety Board, May 2015. [Online; posted 19-May-2015.]
11. Leen G and Heffernan D, Expanding Automotive Electronic Systems, Computer, February 2002;35(1): 88–93.
12. Levinson J, et al., Towards Fully Autonomous Driving: Systems and Algorithms, IEEE Intelligent Vehicles Symposium (IV), July 2011, pp. 163–168.
13. Davies A, Self-Driving Cars Will Make Us Want Fewer Cars, Wired, March 2015. [Online; posted 09-March- 2015].
14. Gao, P, Hensley R, and Zielke A, A road map to the future for the auto industry, McKinsey Quarterly, October 2014. [Online; posted 01-October-2014].
15. Isermann R, Mannale R and Schmitt K. Collision avoidance systems PRORETA: situation analysis and intervention control. Control Engineering Practices, November 2012;20: 1236–1246.
16. Oikawa S, Matsui Y, and Sakurai T. Characteristics of collision damage mitigation braking system for pedestrian protection. International Journal of Automotive Technology, December 2014;15: 1129–1136.

17. Keller CG, Dang T, Frit H, et al. "Active Pedestrian Safety by Automatic Braking and Evasive Steering." IEEE Transactions on Intelligent Transportation Systems, December 2011;12: 1292–1304.
18. Itoh M, Horikome T and Inagaki T. Effectiveness and driver acceptance of a semi-autonomous forward obstacle collision avoidance system. Applied Ergonomics, September. 2013;44: pp. 756–763.
19. Zhao Zhiguo, Zhou Liang Jie, Zhu Qiang, Luo Yugong, and Li Keqiang. A review of essential technologies for collision avoidance assistance systems. Advances in Mechanical Engineering, October 2017; 9(10): 1–15.
20. Schieben A, Griesche S, Hesse T, et al. Evaluation of three different interaction designs for an automatic steering intervention. Transport Research Part F: Traffic Psychology and Behavior, November 2014;27: 238–251.
21. Yu G, Tan D, and Ma F, Analysis and research of issue related to automotive anti-collision system warning/collision algorithm. Journal of Shandong University of Technology (Natural Science Edition), 2014;28(6): 1–5.
22. Sharma A, Zheng Z, Bhaskar A, and Haque MM, Modelling car-following behaviour of connected vehicles with a focus on driver compliance, Transportation Research Part B: Methodological. 2019;126 (12): 256–279.
23. Jiang N, Tian F, Li J, et al., MAN: Mutual attention neural networks model for aspect-level sentiment classification in SIoT. IEEE Internet of Bings Journal. 2019;15(3): 1054–1065.
24. Zhang L, Teng F, Peng Z, et al., Improved vehicle anti-collision warning algorithm based on Berkeley model. Mechanical Science and Technology for Aerospace Engineering. 2018;37(7): 1082–1088
25. Sharma S and Kaushik B, A survey on internet of vehicles: Applications, security issues & solutions, Vehicular Communications. 2019;20(5): 725–729.
26. Automated Driving Systems: A Vision for Safety – NHTSA, https://www.nhtsa.gov/sites/nhtsa.gov/files/documents/13069a-ads2.0_090617_v9a_tag.pdf, 2017.
27. Tigadi A, Gujanatti G, and Gonchi A, Advanced Driver Assistance Systems, International Journal of Engineering Research and General Science. 2016; 4(3).
28. https://www.natlawreview.com/article/dangers-driverless-cars#:~:text=Despite%20claims%20to%20the%20contrary,million%20miles%20for%20regular%20vehicles, 2017.
29. Mo J and Lee Y, Development of a Design Specification and Verification Test Method for Vision-based Oncoming Vehicle Collision Avoidance System. ICCAS. 2017
30. https://mobility.here.com/learn/smart-transportation/vehicle-vehicle-communication
31. https://iprotech.com/industrynews/how-3-cases-involving-self-driving-cars-highlight-ediscovery-and-the-iot/
32. Matsumi R, Raksincharoensak P, and Nagai M, Autonomous Braking Control System for Pedestrian Collision Avoidance by Using Potential Field. IFAC Proceedings Volumes, 2013; vol. 46 (no. 21): pp. 328–334.
33. Ali M, Falcone P, Olsson C, and Sjöberg J, Predictive Prevention of Loss of Vehicle Control for Roadway Departure Avoidance, IEEE Transactions on Intelligent Transportation Systems. 2013;14(1).

34. Ali M, Falcone P, and Sjöberg J, A Predictive Approach to Roadway Departure Prevention, 21st International Symposium on Dynamics of Vehicles on Roads and Tracks, Stockholm, Sweden, 2009.
35. Brännström M, Coelingh E, and Sjöberg J, Model-based Threat Assessment for Avoiding Arbitrary Vehicle Collisions, IEEE Transactions on Intelligent Transportation Systems. September 2010;11(3): 658–669.
36. Ben-Messaoud W, Basset M, Lauffenburger J-P, and Orjuela R, Smooth Obstacle Avoidance Path Planning for Autonomous Vehicles, IEEE International Conference on Vehicular Electronics and Safety (ICVES), September 12–14, 2018
37. Attia R, Daniel J, Lauffenburger J-P, Orjuela R, and Basset M, Reference generation and control strategy for automated vehicle guidance, The IEEE Intelligent Vehicles Symposium (IV), Madrid, Spain, 2012, pp. 389–394.
38. http://www.subaru.com.au/why-subaru
39. Cabrera A, Gowal S, and Martinoli A (eds). A new collision warning system for lead vehicles in rear-end collisions, 2012 IEEE Intelligent Vehicles Symposium (IV). New York, 2012.
40. Mon Y-J and Lin C-M. Supervisory recurrent fuzzy neural network control for vehicle collision avoidance system design. Neural Computing and Applications. 2012;21(8): 2163–2169.
41. Chakraborty I, Tsiotras P, and Lu J (eds). Vehicle posture control through aggressive maneuvering for mitigation of T-bone collisions, 50th IEEE Conference on Decision and Control and European Control, New York, 2011.

Chapter 3

Behavior analysis of broadcast schemes in vehicular accident warning systems against the two-second driving rule

Niaz Chowdhury[*]
Knowledge Media Institute, The Open University
Milton Keynes, United Kingdom

Lewis Mackenzie
Department of Computing Science, University of Glasgow
Glasgow, United Kingdom

CONTENTS

[*] This work is funded by the Scottish Overseas Research Students Award (SORSA) and the University Glasgow College of Science and Engineering Scholarship.

DOI: 10.1201/9781315110905-3

3.1 INTRODUCTION

Broadcast is one of the most popular data dissemination schemes among researchers who build accident warning systems (AWSs) over vehicular ad hoc networks (VANETs). AWSs aim to avoid potential collisions by sending warning messages to neighbor vehicles. The use of the broadcast scheme in existing AWSs is motivated by the highly dynamic nature of VANETs, where nodes move very fast and topology changes frequently. In such an environment, broadcast performs better than multicast or unicast protocols. On the other hand, the two-second rule is a widely accepted approach of defensive driving in many countries, including the United Kingdom, Ireland, and the United States. This rule states that a driver should maintain at least a two-second distance between his or her vehicle and the vehicle ahead of it to avoid possible incidents. This two-second window potentially gives AWSs time to communicate with each other and make decisions about cautioning the drivers.

Although broadcast seems to be an ideal data dissemination scheme for AWSs, it has some drawbacks that can lead to a collapse of the communication system. Broadcast sometimes propels the network toward the verge of breaking down by generating broadcast storms due to excessive transmissions. During a broadcast storm, nodes fail to communicate with neighboring vehicles for some period. If this period is long enough to isolate the host vehicles from its neighbors, it can make AWSs vulnerable. Furthermore, existing AWSs described in the literature have typically taken a generalized approach. Their performance evaluation has often been based on unrealistic simulation and mobility models that do not represent vehicles' movement on roads well. As a result, those systems have never been tested on handling broadcast storms, particularly against the two-second window drivers should maintain when following another vehicle, and therefore a gray area remains unexplored.

This chapter aims to analyze existing AWSs operating with broadcast-based data dissemination schemes to explore their ability to comply with the two-second rule. In doing so, three AWSs from three broad categories of broadcasts—namely *limited-scope, flooding,* and *single-hop*—have been chosen to scrutinize their performances. The chapter then presents a realistic simulation environment and a city mobility model with a flow of vehicles that mimics a real city to evaluate the schemes. This model is then used to analyze and understand the effect of the broadcast storm on the performance of the investigated schemes against the two-second rule.

This chapter's main contribution is the finding that broadcast schemes cannot handle excessive transmissions and often exhibit long transmission delays in such scenarios. This behavior leads to the collapse of the AWS for more than two seconds in many cases. During these phases, vehicles can move along road segments and pass through junctions with little or no ability to communicate with their neighbors. These findings demonstrate that broadcast schemes are not reliable enough to disseminate warnings in AWSs consistently. Therefore, an enhancement on top of these schemes is a necessity to comply with the two-second rule.

The remainder of the chapter is organized as follows: section 3.2 provides the background and the related work; section 3.3 presents the research questions addressed in this chapter; section 3.4 presents the preliminaries, which include an introduction to the investigated protocols followed by a brief introduction to the simulation environment specially constructed for the studies presented later; section 3.5 describes the methodology; and sections 3.6 and 3.7 present the studies.

3.2 BACKGROUND AND RELATED WORK

AWSs are used in VANETs to avoid potential collisions and spread safety notifications among neighboring vehicles.[1] While disseminating data, vehicles sometimes suffer from delays due to excessive transmissions or high network density. As AWSs are cooperative systems, these delays can be crucial at bends, junctions, or even on straight roads.[2]

One of the key causes of these delays is the data dissemination approach. Depending on the type of schemes used, delays can vary. Nevertheless, the process of data dissemination in VANETs is challenging because of their dynamics,[3-5] and an added layer of difficulty is introduced when the VANET is the basis of a real-time AWS. The review presented by Chowdhury et al. in[2] shows that existing proposals for AWSs that address the problem related to collision and accident avoidance mostly rely on broadcasting for the data dissemination scheme to cope with this highly dynamic nature of VANETs.

It is also notable that broadcast schemes used in various AWSs are not identical. Those schemes can be divided into three broad categories: limited-scope, flooding, and single-hop broadcast. *Limited-scope* can be defined as a dissemination technique that sends data to all nodes within a defined boundary, such as a certain number of hops or a small geographical region. The definition of *flooding* can be as simple as the technique of sending data to all connected nodes. The last category, *single-hop broadcast*, disseminates data only to neighbors located within a one-hop distance.

The performance varies depending on the type of their categories. For example, flooding provides the best coverage but injects a lot of redundant data into the network. In contrast, single-hop broadcast offers the least coverage but shows less redundancy in terms of data overhead. It is no surprise that a large number of existing protocols use limited-scope, because it allows for a balance between coverage and redundancy. Despite their different characteristics, all three categories follow the basic principles of stochastic (random) broadcasting.

However, broadcast is infamous for creating *storms* in the network that can isolate a node from its neighbors for some period. A broadcast storm can be explained in brief as follows: In an ad hoc network, when a node broadcasts a packet to its one-hop neighbors, it usually receives that packet almost simultaneously. An imperative to rebroadcast that packet instantly results in all receiver nodes trying to get channel access that incurs collision. To avoid such a situation, stochastic broadcast protocols in ad hoc networks may deliberately insert a small but random delay called *random assessment delay (RAD)* in the scheduling of data delivery from the network layer to the link layer so that neighboring nodes rebroadcast data at different times.[6, 7] However, when the number of rebroadcasts considerably increases, contention to get channel access becomes fierce, and the RAD alone cannot prevent nodes from getting involved in collisions. If the network continues to experience such behavior, it results in a broadcast storm.[8-10]

Several previous studies showed how this phenomenon affects various networks and applications while operating with broadcast schemes.

Ni et al. in,[11] through numerical analysis, depicts how a broadcast storm is generated in Institute of Electrical and Electronics Engineers (IEEE) 802.11 multi-hop wireless networks and argues that as the number of neighbor nodes increases, throughput performance falls. Choi et al. in[12] and Tseng et al. in[13] investigate the problem from the ad hoc network's perspective and identify that rebroadcast is the key factor in creating broadcast storms that lead to redundancy, contention, and collision in the network. Wisitpongphan et al. in[14] and Martinez et al. in[15] reinvestigate the issue from the VANET's perspective and conclude that (i) high link load causes high contention in the network, resulting in packet loss, and (ii) low packet penetration causes a long delay. However, previous work has not investigated the behavior of broadcast schemes with a real-time system like an AWS, a gap this chapter aims to fill.

3.3 TWO-SECOND RULE AND RESEARCH QUESTIONS

The investigated AWSs in the studies presented in this chapter will be evaluated as to how they comply with the two-second rule. As previously introduced, the two-second rule is considered a rule of thumb by which a driver can maintain a safe distance between the driver's vehicle and the vehicle ahead of it on the road. The rule suggests that a driver should ideally stay at least two seconds behind any moving vehicle that the driver is following. Road safety authorities in the United Kingdom, Ireland, and the United States promote this rule.[16–18] The two-second rule is used in this chapter to establish a benchmark delay that defines the maximum period an AWS can potentially be isolated from other ASWs before a dangerous condition is created. In reality, this isolation occurs when warnings await in the delivery queue but the underlying MAC layer does not get access to the medium due to competition. Therefore, this period is named the *maximum tolerable queuing delay (MTQD)*.[19]

This chapter addresses two research questions. The first focuses on the two second-rule. It determines whether exiting broadcast-based AWSs can comply with this rule by maintaining a queuing delay below the MTQD threshold. Therefore, the first step of this research is to analyze the broadcast schemes to find the delay they experience in different situations, particularly in the presence of broadcast storms. The second research question addresses whether broadcast-based AWSs can detect accidents, primarily while operating with a queuing delay longer than the MTQD. This research question will be tested against intentionally created potential accidents to observe how the AWS schemes respond to those situations.

3.4 PRELIMINARIES

This section discusses the knowledge and information required to understand the studies presented in this chapter. It includes a description of the investigated protocols and an introduction to the simulation environment built as a part of these studies.

3.4.1 Investigated accident warning systems

Upon reviewing all available AWSs and categorizing them into the previously mentioned three broad categories, we chose three AWSs, one from each category. Our aim was to pick the best-performing system from each category, with the exception of the single-hop category, which had only one choice. Brief descriptions of those three investigated AWSs are presented next.

3.4.1.1 VANET solution to prevent car accidents (VSPCA)

VSPCA typifies the use of flooding. This AWS is more sophisticated than other members of its category in terms of both coverage and data dissemination. It generates messages and warnings for a diverse range of scenarios, including collision avoidance messages and event-driven post-incident warnings.[20]

3.4.1.2 Reliable broadcasting of safety messages in VANET (RBSM)

RBSM represents the limited-scope broadcast category. It tries to establish a balance between the importance of the reachability of a warning and the consequences the network suffers due to flooding. RBSM introduces two critical elements in its design. First, it tries to control flooding by using a parameter that keeps track of the maximum number of times a message can be forwarded. Second, it uses beacon messages to allow any given vehicle to learn about its neighborhood. The first element enables senders to control the scope of a particular warning, and therefore turns flooding into a limited-scope broadcast,[21] and the second element helps keep track of neighbors continuously.

3.4.1.3 IEEE 802.11p

IEEE 802.11p is a single-hop broadcast protocol.[22] No other system currently uses single-hop message delivery other than the built-in beacon frame embedded here (802.11p). Although the general beaconing rate is twice a second, it is altered in this study to explore its performance further by placing a simple network layer on top of it.

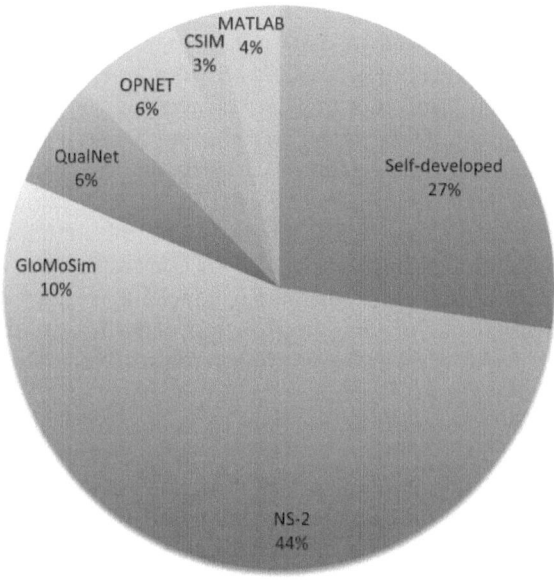

Figure 3.1 Simulator usage from the MobiHoc survey.

3.4.2 Simulator

Simulators are considered a valuable tool for research when real-life trials are not possible before completing the system's physical design. A simulated study also provides the opportunity to evaluate more functionalities than a real trial in many cases. Nevertheless, finding a realistic simulation environment is difficult because mainstream simulators such as NS2, NS3, OMNET++, OPNET, etc.[23–25] often come with predefined mobility models, obstacle models, protocols, and parameters that require a lot of effort to customize. Calibrating these features in line with a particular experiment's specific requirements is not always possible. For example, a mobility model for nodes with various velocities is a standard feature in most available simulators. But this is typically implemented as some pattern-based or randomly generated movements that are unrealistic and unsuitable for a vehicular environment.[26] Some open-source simulators have been developed recently (such as Veins[27]) that provide custom-built mobility models, but development without sufficient technical support and community contribution often limit their usefulness.

A third option is for researchers to develop their own custom-built, self-deployed simulator. A study reveals that between 2000 and 2005 in MobiHoc (Fig 3.1), one of the most prestigious mobile network conferences, 27 percent of authors used self-developed simulators.[28] Although NS2 was the most popular simulator at the time of that study, the self-developed simulator was still the second most popular choice by a large margin.

These self-deployed simulators are highly customizable and allow including specific functionalities in the experiments. The current work borrowed only the medium access control (MAC) layer from an existing simulator and developed a network layer and application layer (as shown in Figure 3.2) along with a realistic mobility model on top of it.

We chose an open-source simulator called *Pamvotis* to provide the support of the MAC layer. It offers various IEEE 802.11 protocols, and we opted for its *p* variant. Pamvotis is a lightweight and highly customizable Java-based simulator that has a provision for connecting any mobility and radio model from outside without the developer having to touch its internal functionalities.[29] While building on top of Pamvotis, this provision ensures that the MAC layer's performance remains unaltered.

We implemented the two network layer protocols and AWSs, VSPCA and RBSM, using Java based on the description provided in the respective

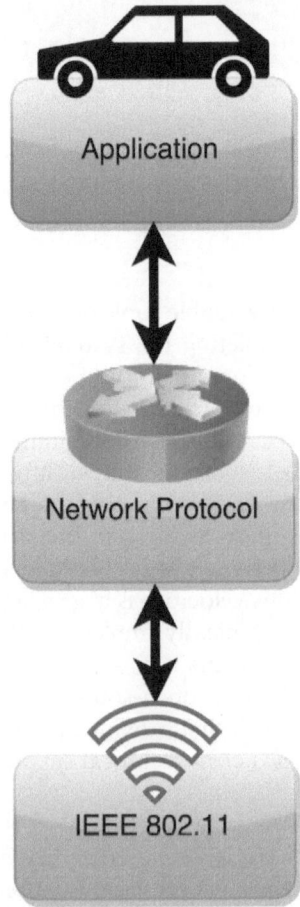

Figure 3.2 The stack view of the protocols in the custom-built simulator used in this study.

publications. We also developed an application layer protocol using Java that generates periodic warnings at a defined rate. We developed a simple network protocol for the single-hop broadcast protocol to act as an interface between the application layer and MAC layer. The protocol's operation is limited to passing warnings between its top and bottom layers. In addition, after considering several radio models that attempt to create vehicular environments, such as described in,[30–32] we implemented a model proposed by Mahajan et al. in[33] and integrated it with the simulator.

3.4.3 Mobility model

Mobility models are used to evaluate the performance of wireless protocols and algorithms by simulating them in an environment where the nodes are in motion. There are two possible mobility model types found in the literature: *trace* and *synthetic* models.[34] Trace models are based on real nodes' movement, while synthetic models follow a mathematical formula that artificially approximates such motion. In the context of AWSs, the trace model is not the right choice, since it cannot be expected to generate possible accident scenarios. Although synthetic models are often considered unrealistic, in the current experiment, they can produce approximately realistic movement if some auxiliary information is fed into the real environment. Musolesi et al. in[35] named such hybrid schemes "synthetic model(s) starting from real traces." This study develops a model of this type called the *Glasgow Mobility Model*.

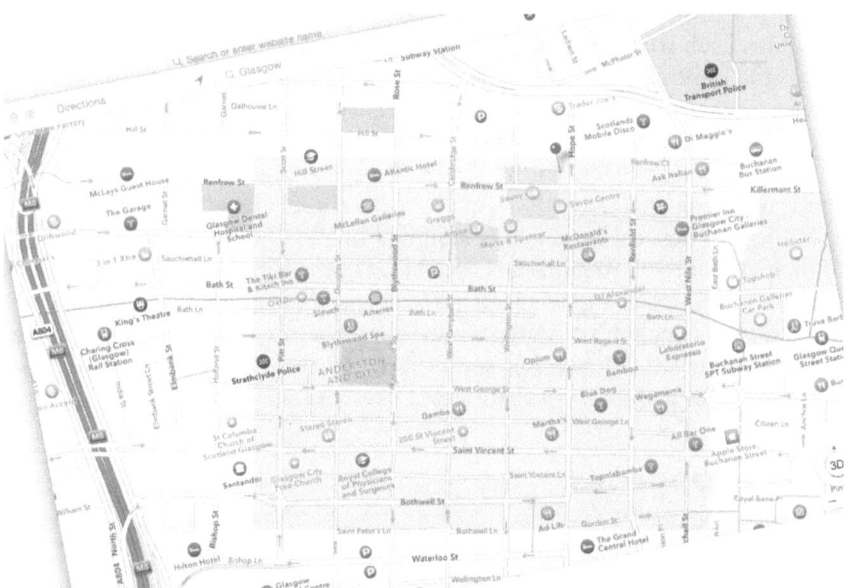

Figure 3.3 Fragment of the Glasgow City Center used in the simulation of this study.

Table 3.1 Vehicle Densities at Glasgow City Center

Density (vehicles/min)	Density Type	Streets
10	Low	Douglas St., West Campbell St., and Wellington St.
20	Moderate	Bothwell St., Pit St., and Blythswood St.
30	Moderate	Bath St., West Regent St., West George St., and West Nile St.
40	High	Sauchiehall St., Renfrew St., and Saint Vincent St.
50	High	Hope St. and Renfield St.

The model uses a real location, Glasgow City Center, as the designed environment's simulation arena. This location is an excellent example of a grid-road scenario where roads run from east to west and north to south, setting up an ideal urban environment for testing AWSs. Figure 3.3 shows Glasgow City Center on an online map. The highlighted area is the busiest part of the center and the active experimental region in this simulation. A noticeable feature on the map is the direction of the movement of the vehicles. All roads in the center area are unidirectional, as shown on the map via arrows. While developing the Glasgow Mobility Model, we strictly following these directions.

The model also uses realistic vehicle density on the roads. To achieve this goal, we measured the peak-hour vehicle movement by visiting the location on five consecutive working days of a typical week. Table 3.1 shows streets and their average density, with data rounded to the nearest whole number.

Finally, each road is assigned a *source* and a *sink*. Sources generate vehicles at the appropriate rate, starting to move from an initial trace but following a formula that leads them towards the sink. When a sink absorbs a vehicle, it immediately passes it on to the corresponding sink so that the vehicle can be sent back on the street. During this exchange, all the communication parameters of the vehicles remain unchanged.

While moving, each vehicle is assigned one of the following velocities: 10 mph, 30 mph, or 50 mph. It is noted that in the United Kingdom, 30 mph is the maximum legitimate speed a driver can maintain in a city area. The speed domain includes a slower speed than the maximum allowed and a faster speed that is illegal, but sometimes drivers reach that limit. In this study, vehicles cannot vary their speed over time.

3.5 METHODOLOGY

This section discusses the methodology of the studies, the setups, and the metrics used to make the observations. Overall, this section establishes the basis of the analysis that will be presented later in sections 3.6 and 3.7.

3.5.1 Simulation environment

The studies presented in this chapter aim to analyze the performance of AWSs by focusing on the effect of excessive transmissions made by the vehicles. To achieve this goal, the custom-built simulator described earlier is used. A closely imitated real-life motor traffic setting of Glasgow City Center representing the single carriageway scenario in the urban layout is constructed. This environment assists in simulating and observing the behavior of the experimental protocols with a broader range of freedom. First, it allows populating vehicles from the source in a particular order to measure specific aspects concerning incidents. Second, it helps to monitor each street and junction separately to have a more comprehensive understanding. And third, it allows generation and dissemination of warnings as per the need of the study.

3.5.2 Performance metrics

The studies presented in this chapter use performance metrics from both the network layer and the application layer. At the network layer, three metrics are evaluated: i) *rate of collision*, a metric that expresses the number of transmission collisions encountered by a node (i.e., vehicle) every second during the simulation time; ii) *queuing delay*, a metric that expresses (in milliseconds) on average how long a warning needs to wait at the respective transmission queue from the time of its generation; and iii) *jitter*, a metric that expresses the standard deviation of the queuing delays for a vehicle.

At the application layer, five metrics are evaluated. These involve three types of situations that need to be explained before the metrics. The first situation is called *potential accident* and is defined as follows: If a vehicle, either while moving straight or at a junction, collides with another vehicle in such a situation that at least one of them failed to receive a warning before the collision, then that situation is called a potential accident. The second situation is called *blind movement* and is defined as a situation in which a vehicle passes a junction with a transmission queuing delay larger than the MTQD. The third situation is called *safe movement* and incorporates the transmission queuing delay smaller than the MTQD.

The first three metrics are evaluated at junctions and do not count how many times they occur; instead, they count how many junctions they appear in. These metrics are as follows: i) *accident-at-junction*, a metric that expresses a potential accident occurred at a particular intersection; ii) *blind-move*, a metric that shows a specific junction has blind movement; and iii) *safe-move*, a metric that indicates a particular junction has safe movement. If an intersection encounters a potential accident, it is counted towards accident-at-junction. However, if a crossing encounters no possible accident but at least one blind movement, it will be counted towards blind-move. All other junctions will be counted towards safe-move. Two other metrics are evaluated that give a total number of *potential accident undetected* and *potential accident detected* during the simulation.

3.5.3 Experimental setup

This chapter reviews two studies that have been conducted, each having two versions: *moderate* and *heavy*. The moderate version examines the network with a moderate probe of either data traffic load or network density, while the heavy version shakes the network with a probing that could well be considered "not normal" in general.

The first study investigates the effect of data traffic load on the performance metrics. This study varied data traffic load between 1 to 25 warnings/sec to inspect how performance metrics react to this variation. Based on the reciprocation received from the performance metrics, this study finds a reasonable warning generation rate for what queuing delay stays below the MTQD. This rate will be later used in the second set of studies as the moderate data traffic load. No application layer performance metrics are evaluated in this study, and the vehicles' initial assignment on the streets does not have a specific order.

The second study investigates the effect of network density on the performance metrics of the experimental AWSs. This study has two objectives as it evaluates both network and application layer metrics. The network layer component examines a warning's transmission collisions in the network and subsequently inspects prolonged queuing delays and their respective jitters. Collisions indicate how stiff the competition is to get access to the medium, while delay reveals the consequences of that competition. Associated jitter sums up the study by demonstrating the variation of the delays.

At the application layer, it has been determined how many junctions in the Glasgow City Center's experimental segments shown in Figure 3.2 have blind-move and accident-at-junction. The number of potential accidents undetected at junctions and on straight streets during the simulation is also evaluated. Vehicles are generated in a particular order at the sources to better understand the results of this part of the study. It is assumed for simplicity that there are two lanes on each street and three fixed velocities available for vehicles. Instead of randomly assigning lanes and velocities, this is intentionally conducted so that a vehicle with a higher speed cannot follow a vehicle with a lower speed in the same lane, to avoid any collision. However, this rule is only broken for the two highest-velocity vehicles placed at the very back of the street. These two vehicles pass through slower vehicles in front of them as the simulation progresses. With this setup's help, potential accidents are deliberately created to evaluate how many of such accidents go undetected when vehicles operate with various AWSs.

In this study, streets are grouped based on their network density. For example, all streets having 50 vehicles/min density are grouped by averaging their performance metrics and so on. Street names could have been used to prepare the graphs, but that would not demonstrate the context of the results, which is network density in this case.

3.6 THE EFFECT OF DATA TRAFFIC LOAD

The study of the effect of data traffic load will be twofold, with two different network densities. The first network density is named *moderate* because it simulates with 5 vehicles/min network density, while the second density is called *heavy* and uses 15 vehicles/min network density. The simulation will run for 60 seconds, and an average of 20 trials is taken to prepare the graphs.

3.6.1 Network density: Moderate

The study of the effect of data traffic load in moderate network density varies data traffic load from 1 to 25 in order of 1, 5, 10, 15, 20, and 25 warnings/sec and observes the following performance metrics.

3.6.1.1 Rate of collision

It has been previously noted in the literature that contention for medium access occurs in wireless networks due to excessive rebroadcast.[12, 13] This study further investigates the issue in the context of AWSs. The top-left chart in Figure 3.4 demonstrates the effect of traffic load on the collision rate in moderate network density. It shows that collisions occur more frequently as data traffic load increases in the network and follows a sharp rise in the load generation rate's growth. Notably, RBSM, which uses limited-scope broadcast, and VSPCA, which uses flooding, experience this sharp increase at an early stage, and later single-hop (IEEE 802.11p) follows the same pattern.

The number of collisions, however, is not similar for all three AWSs. The most notable attribute observed in this study is the behavior of 802.11p, which encounters fewer collisions compared to the other two systems when the data traffic load stays below 15 warnings/sec. However, when data traffic load surpasses this limit, 802.11p encounters more collisions than RBSM and VSPCA despite forwarding warnings no more than a single-hop distance. This study also finds that the network does not suffer from contention when the data traffic load is 5 warnings/sec or less. With the data traffic load increasing further, AWSs are more likely to encounter excessive collisions that result in the creation of a broadcast storm in the network.

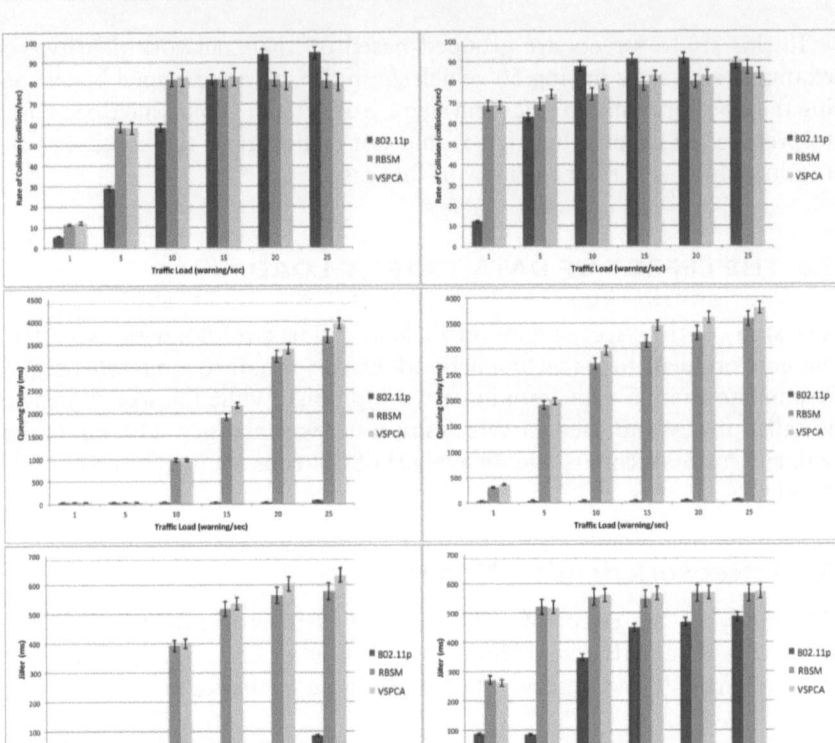

Figure 3.4 The effect of data traffic load on transmission collisions (top), queuing delays (middle), and jitters (bottom) in two network densities: *moderate* in the left column and *heavy* in the right column.

3.6.1.2 Queuing delay

The middle-left chart in Figure 3.4 shows that the queuing delay of warnings increases rapidly as the data traffic load grows. It happens when multiple vehicles attempt to send warnings via the wireless medium simultaneously, and all but one sender needs to back off to avoid potential interference.[22] As a result of this behavior, warnings waiting in the queue are required to stay there until the sender vehicle receives access to the medium.

From this chart, it is also evident that RBSM and VSPCA are almost identical at the moderate network density, because at a network density of 5 vehicles/min, these two systems perform similarly. As data traffic load grows, both face a sharp rise in their queuing delay, and at the peak of 25 warnings/sec, this delay hits nearly 4 sec, a delay almost twice the MTQD, for each AWS. 802.11p, however, performs better because it does not rebroadcast warnings. It remains unaffected and maintains a consistent and minimal queuing delay throughout.

This observation indicates that 5 warnings/sec is a reasonable warning generation rate, as the queuing delay of all three protocols for that rate is minimal. However, RBSM and VSPCA stay within the borderline of the MTQD when this rate is 15 warnings/sec.

3.6.1.3 Jitter

The bottom-left chart in Figure 3.4 shows the jitter in the queuing delay. As this metric reveals the variations in the delay, a large value of jitter indicates an inconsistency in the queuing delay. It is highly likely that this inconsistency is generated by the contention for accessing the medium.

From this chart, it is evident that the jitter increases as the delay grows. It further emphasizes that with the growing contention in the network, the delay becomes inconsistent. This study reveals that although jitter is consistent for 802.11p throughout the simulation, it varies between 400 millisecond (ms) and 600 ms when the data traffic load is more than 5 warnings/sec for RBSM and VSPCA. With the other two metrics, collision and queuing delay, This study also finds that 5 warnings/sec is a reasonable warning generation rate in moderate network density.

3.6.2 Network density: Heavy

In the study of the effect of data traffic load in heavy network density, the network density is 15 vehicles/min and the three performance metrics evaluated in the previous section are used again.

3.6.2.1 Rate of collision

The top-right chart of Figure 3.4 shows that all three systems exhibit a sharp increase, as it does in the case of moderate network density, but it now starts earlier. 802.11p initially maintains a slower growth but later surpasses the other two systems. However, it is notable that despite having a network density of three times the previous study, the rate of collision does not change. This indicates that although network density growth plays an essential part in creating contention in the network, contention does not strictly follow the growth by getting increased proportionally.

3.6.2.2 Queuing delay

The middle-right chart of Figure 3.4 demonstrates that the queuing delay in heavy network density is sharply higher than the previous study, where more than 5 warnings/sec rate was required to have such growth. It is evident from this study that performance of limited-scope-based RBSM and flooding-based VSPCA are almost identical. At the peak of the data traffic load of 25 warnings/sec, this delay exceeds 3 sec for each system.

802.11p, however, performs better and exhibits identical behavior to the case of moderate network density. It is also evident from this study that the performance of limited-scope-based RBSM and flooding-based VSPCA are almost similar, and at the peak of the traffic load with 25 warnings/sec, this delay hits above 3 sec for each system.

This study reconfirms that 5 warnings/sec is a reasonable warning generation rate. Even though the network density is 15 vehicles/min, both RBSM and VSCPA stay within the MTQD.

3.6.2.3 Jitter

The bottom-right chart of Figure 3.4 demonstrates the jitter in heavy network density. Note that, unlike the moderate network density case, here the jitter of 802.11p also increases. This indicates that if the data traffic load in a dense environment increases, the warning delivery time will vary significantly even if restricted to a single-hop boundary. The figure shows RBSM and VSPCA demonstrate high jitter from 5 warnings/sec traffic load onwards. The 802.11p protocol still performs better and remains consistent concerning warning delivery time at or below 5 warnings/sec.

3.6.3 Discussion

The study of the effect of data traffic load is one of the two studies performed in this chapter to address the first research question mentioned earlier in section 3.3: whether the existing broadcast-based AWSs can comply with the two-second rule and are capable of maintaining a queuing delay below the MTQD threshold.

The current study diagnoses how AWSs perform in response to data load and shows that collisions occur more frequently as data traffic load increases in the network and follows a sharp rise with the growth of the warning generation rate. The study also identifies that the network does not suffer from contention when the data traffic load is 5 warnings/sec or below with network density of 5 vehicles/min. However, as data traffic load increases further, AWSs are more likely to encounter excessive collisions that result in the creation of a broadcast storm in the network. This study then relates collision with the queuing delay and identifies that as the likelihood of having a broadcast storm increases, the queuing delay increases too. It strengthens this discovery by showing a queuing delay in 15 vehicles/min network density (three times larger than the previous load) that sharply grows compared to the low network density. Finally, by analyzing the jitter, it demonstrates that if the data traffic load in a dense environment increases, the warning delivery time will vary significantly, even if the dissemination is restricted to a single-hop boundary.

Based on the preceding summary, this study partially answers the first research question by confirming that existing AWSs cannot comply with the MTQD threshold with growing data traffic loads on the network. It also finds that 5 warnings/sec is a reasonable warning generation rate. In both 5 and 15 vehicles/min network densities, AWSs maintain a queuing delay below the MTQD while operating with this rate.

3.7 THE EFFECT OF NETWORK DENSITY

The study of the effect of network density is performed by varying the density of vehicles while keeping other parameters fixed. This study is twofold, with two different data traffic loads. The first data traffic load is named *moderate* because it uses a reasonable probe of 5 warnings/sec recommended in the previous study. The second data traffic load is called *heavy* and uses a 15 warnings/sec data traffic load. This study is conducted to evaluate all six performance metrics mentioned in section 3.5.2 and the simulation runs for 60 sec.

3.7.1 Data traffic load: Moderate

In this phase of the study, the network density is determined by the vehicle rate on streets shown in Table 3.1. Five values of network density, 10, 20, 30, 40, and 50 vehicles/min, are used in this study. An average of 20 trials is taken for each metric for preparing the graphs.

3.7.1.1 Rate of collision

The top-left chart of Figure 3.5 shows the collision rate (transmission collision) in the network. It is evident from the results that when more vehicles enter into the same enclosed area, competition for accessing the medium gets stiffer and streets with more vehicles tend to have more collisions.

This behavior suggests that as network density increases, the number of rebroadcasts of warning messages significantly increases in the network, ultimately resulting in a broadcast storm.

The results show that both RBSM and VSPCA with limited-scope and flooding techniques, respectively, encounter more than 25 collisions when network density is 10 vehicles/min. The IEEE 802.11p beacon, however, encounters fewer collisions because it does not rebroadcast warning message. However, as network density increases, the broadcast storm looks severe, and all three systems see a more than 40 percent rise in collisions.

3.7.1.2 Queuing delay

The results in the middle-left chart of Figure 3.5 demonstrate the network density's effect on the queuing delay of warnings while operating with moderate data traffic load. It clearly shows that, like the previous

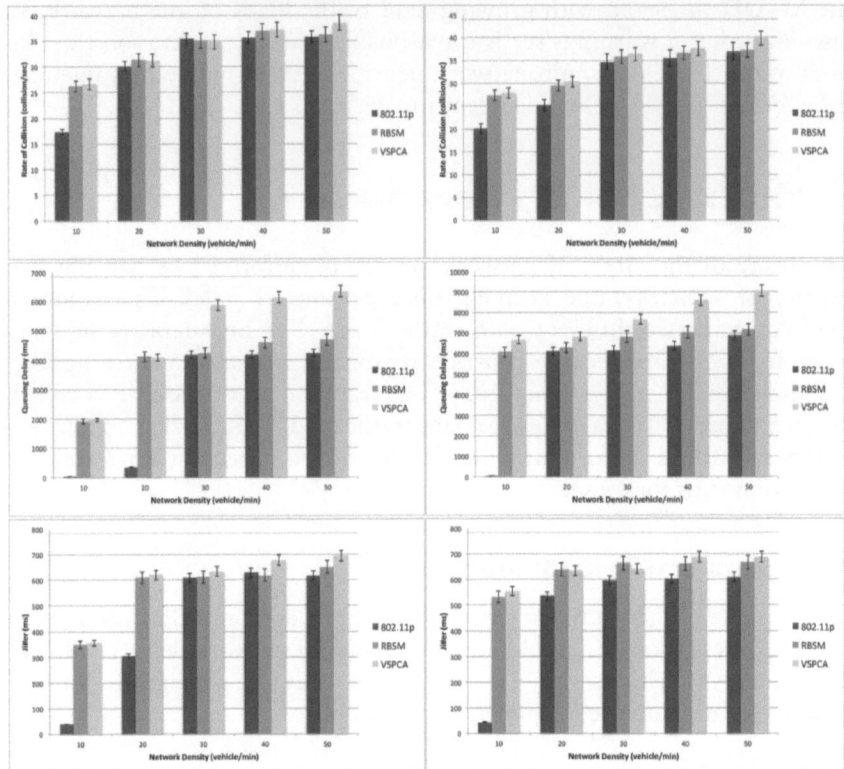

Figure 3.5 The effect of network density on transmission collisions (top), queuing delays (middle), and jitters (bottom) in two data traffic loads: *moderate* in the left column and *heavy* in the right column.

results, there exists a relation between queuing delay and network density; and with more vehicles entering on a street, competition for accessing the medium gets stiffer. It also demonstrates that collision is a cause generated by the broadcast storm, and queuing delay is its consequence.

VSPCA exhibits a sharp rise that begins with a delay just below 2 seconds but becomes double as the network density doubles. Subsequently, this delay touches 6 seconds, three times the MTQD, when network densities are between 30 and 50 vehicles/min. This behavior can be explained by realizing that as the number of vehicles increases in the network, the number of rebroadcasts also increases, resulting in a broadcast storm and subsequently long queuing delay. As RBSM only forwards warnings up to a five-hop distance and generates fewer rebroadcasts than flooding, it exhibits less queuing delay than VSPCA. When network density is not high, both of these systems perform almost identically. This is because the distance between any two vehicles rarely could have been more than five hops during this phase. However, when the number of vehicles increases, the possibility of obtaining a path longer than five hops becomes prominent and performance of limited-scope-based RBSM deviates from its flooding counterpart. It is also noted that IEEE 802.11p exhibits a small queuing delay for network density of 20 vehicles/min or less. Nevertheless, it shows a very sharp rise afterwards and continues with a queuing delay similar to that of RBSM.

3.7.1.3 Jitter

The bottom-left chart of Figure 3.5 shows the average jitter at different network densities. As with the previous two performance metrics, it also demonstrates a sharp growth. It is evident from the results that although jitter initially increases sharply, as network density increases, it becomes stable. This is because this jitter is the standard deviation of the queuing delay demonstrating how those delays fluctuate. As the broadcast storm becomes severe, delays tend to fluctuate less.

The results show that with network density of 10 vehicles/min, jitter is nearly 350 ms for RBSM and VSPCA; and as network density gets doubled, it climbs to more than 600 ms. The IEEE 802.11p beacon maintains a flat rise compared to the other protocols but eventually matches their performance.

3.7.1.4 Junctions

There are a total of 44 junctions in the simulation, as shown in Figure 3.6. When vehicles move through these junctions, three situations can occur: i) accident-at-junction; ii) blind-move, and iii) safe-move.

The top-left chart in Figure 3.6 demonstrates that flooding-based VSPCA encountered more accidents than any other protocol, while RBSM encountered more blind-moves. Results shown in sections 3.7.1.1, 3.7.1.2, and 3.7.1.3 explain this behavior. As these two AWSs experienced long delays in high network densities, they encountered more potential accidents or had numerous blind-moves at junctions. However, it is notable that despite showing better performance than the other two systems in terms of delay, the IEEE 802.11p beacon also encountered a significant number of accidents at junctions. The reason is that, with its more limited coverage, vehicles often fail to propagate warnings to the appropriate recipients.

3.7.1.5 Potential accident undetected

Streets are grouped based on their densities in the simulation. The number of potential accidents undetected in those groups are then recorded. As previously mentioned in section 3.5, vehicles are placed on the street at the beginning of the simulation. Two vehicles with 50 mph velocity from the very back hit slower vehicles while moving forward. These deliberate

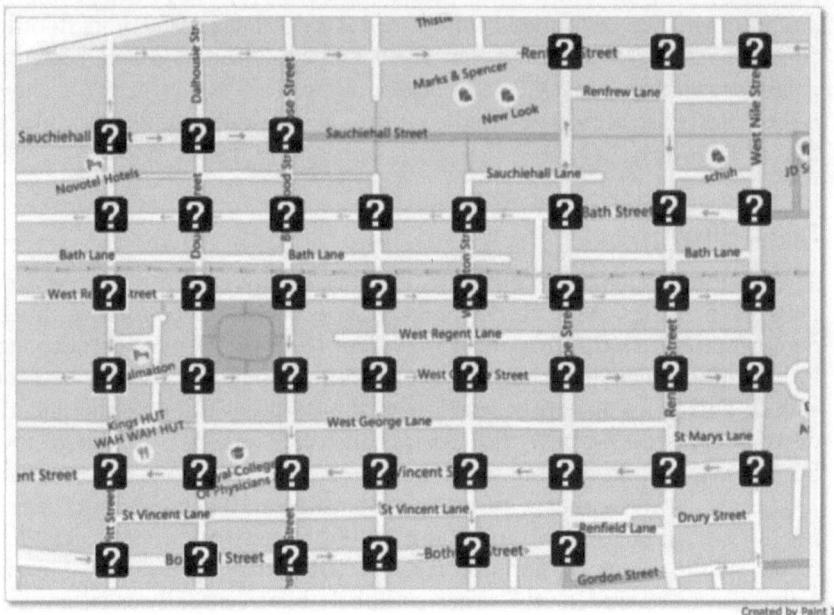

Created by Paint X

Figure 3.6 Junctions used in the studies to evaluate accident-at-junction, blind-move, and safe-move.

actions ensure that accidents occur in the simulation and therefore pave the way for testing the AWSs' effectiveness.

The middle-left chart of Figure 3.7 shows that VSPCA has more potential accidents undetected than any other AWSs, while RBSM and 802.11p perform almost identically except for the highest network density. The performance observed here is quite simple to understand: when vehicles experience long queuing delays, they fail to disseminate warning messages on time, which ultimately makes their AWSs vulnerable. It is essential to realize that despite having a better queuing delay than RBSM for the network density 20 vehicles/min or less, 802.11p did not perform well. Quicker delivery of the warning messages is vital to spread them to the right audience—a fact that cannot be overlooked. Having covered only one-hop distance neighbors, 802.11p fails to extend the warnings to all required vehicles.

It is also notable that in high network densities, the number of potential accidents undetected remains between 35 and 45, while in the low network densities it is just around 5. This is understandable, because a broadcast storm influences this performance unfavorably by pushing the queuing delay towards, or more than, the MTQD. To improve the AWSs' performance, a countermeasure is required that prevents broadcast storms from being generated in the network, hence keeping the queuing delay below the MTQD.

3.7.1.6 Potential accident detected

The number of potential accidents undetected gives a picture of how badly the systems perform. However, it does not account for how many accidents are being detected and a complete picture cannot be observed without exploring both metrics.

The bottom-left chart of Figure 3.7 demonstrates the number of potential accident detected during the simulation. While VSPCA encounters nearly 45 accidents in the high densities, it only detects around 10 potential accidents. So roughly, for every five potential accidents, VSPCA is capable of detecting only one. The other two AWSs perform relatively better, as they can detect one out of three accidents.

3.7.2 Data traffic load: Heavy

In this phase of the study, a traffic load of 15 warnings/sec is used to observe how performance metrics respond to such a heavy data traffic load. As with the previous study, the network density is determined by vehicles' rate on streets, as shown in Table 3.1. Five network density values, 10, 20, 30, 40, and 50 vehicles/min, are used. An average of 20 trials is taken for each metric.

3.7.2.1 Rate of collision

The top-right chart of Figure 3.5 shows the number of collisions in the network in the presence of heavy data traffic load. If this result is compared with the previous results described in its moderate counterpart, a slight increase in the number of collisions can be observed for similar network densities. However, this behavior is expected, because with more warnings entering into the network, collisions are likely to increase.

This study demonstrates that both RBSM and VSPCA encounter just over 25 collisions initially but later settle between 35 and 40 collisions. The IEEE 802.11p beacon encounters fewer collisions because it does not rebroadcast warning messages. However, as network density increases, all three systems exhibit a rise in the number of collisions.

3.7.2.2 Queuing delay

The results in the middle-right chart of Figure 3.5 demonstrate network density's effect on the queuing delay of warnings while operating with heavy data traffic load. The previous section shows that a large number of collisions in the network is also an indicator of fierce competition for medium access. This figure confirms that large delays are outcomes of the broadcast storms having experienced heavy transmission collisions by the vehicles. It is noteworthy that due to the additional warnings in the network, this delay gets prolonged compared to its moderate counterpart performance presented earlier.

This figure shows that all three systems exhibit long delays well above the MTQD from the beginning. Although RBSM and IEEE 802.11p show stable performance, VSPCA demonstrates a growth throughout the simulation period. This behavior can be explained by realizing that when the number of vehicles and the number of data traffic loads increase, fierce competition occurs in the network to get access to the medium. It results in a very long queuing delay for sender vehicles.

3.7.2.3 Jitter

The bottom-right chart of Figure 3.5 shows the average jitter when the data traffic load is 15 warnings/sec. Comparing this result to the result presented in the middle-right chart of Figure 3.5 provides a complete picture showing the delays to be very stable. For example, when RBSM experiences a 9-sec queuing delay, its corresponding jitter is only 700 ms, which is the same as presented in Figure 3.5 (bottom left) for 5 warnings/sec data traffic load. This behavior can be explained as follows: if the data traffic load increases, vehicles encounter more contention and, hence, hardly get access to the medium.

The result shows that with network density of 10 vehicles/min, jitter is more than 500 ms for VSPCA and RBSM, but with 40 vehicles/min or more, it climbs to near 700 ms. The 802.11p beacon also maintains a jitter between 500 and 600 ms except for with a 10 vehicles/min network density.

3.7.2.4 Junctions

As previously mentioned, the simulation has a total of 44 junctions and vehicles may experience *blind-movement* or *safe-movement* while passing through these junctions or encounter a *potential accident*. In the presence of a heavy data traffic load, the queuing delay sharply rises from the very beginning, deviating from the performance demonstrated for a moderate data traffic load. This change in queuing delay significantly affects the statistics measured for the circumstances recorded at junctions.

The top-right chart of Figure 3.7 demonstrates that all AWSs encounter almost the same number of accidents as they do for moderate traffic load. However, the safe-move drastically decreases while the blind-move increases for each of them. The long queuing delays influence this behavior change even in low network densities.

3.7.2.5 Potential accident undetected

As mentioned earlier, the streets are grouped based on their densities. The number of accidents undetected in those groups is recorded during the simulation. Section 3.5 describes the placement of the vehicles and how they are released at the beginning of the simulation on the streets.

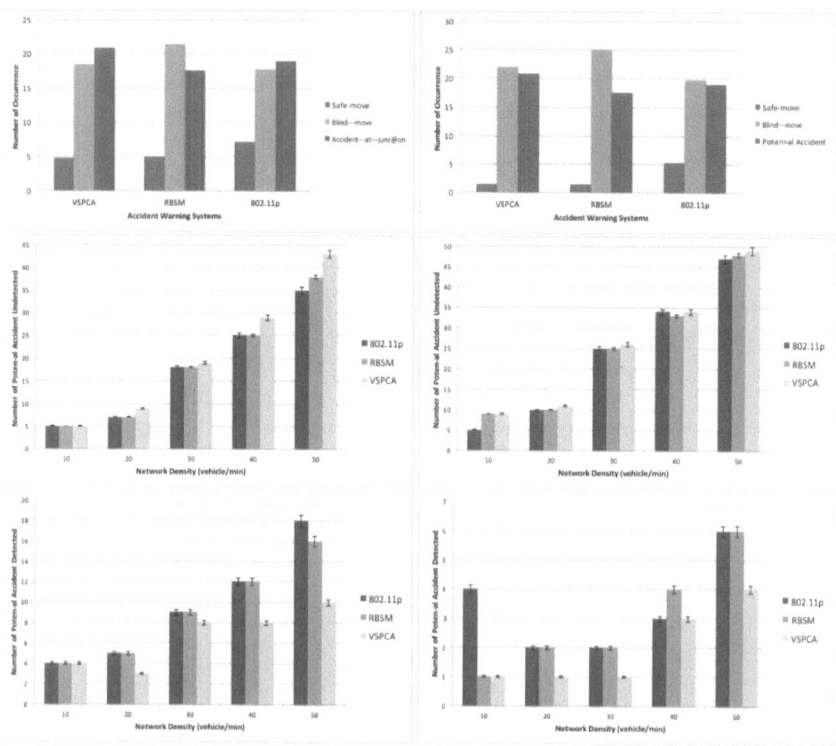

Figure 3.7 Movements at the junctions (top), the number of potential accidents unde-tected (middle), and the number of potential accidents detected (bottom) in *moderate* data traffic load.

The middle-right chart of Figure 3.7 shows all AWSs have more potential accidents undetected than they have in moderate data traffic load. Among them, however, VSPCA has more undetected potential accidents than any other system. On the other side, RBSM and 802.11p perform almost identi-cally except for the lowest network density. The performance observed here further confirms that when vehicles experience long queuing delays, signifi-cantly longer than the MTQD, they fail to disseminate warning messages on time, which ultimately makes their AWSs vulnerable.

3.7.2.6 Potential accident detected

The bottom-right chart of Figure 3.7 demonstrates the number of poten-tial accidents detected during the simulation. While VSPCA has nearly 50 undetected potential accidents in the high network densities, it has only around 4 potential accidents detected. This means that, roughly, for every 12 potential accidents, VSPCA can detect only one when the warning dissemination rate (i.e., data traffic rate) is very high, in this case, 15 warnings/sec. The other two systems also fail to demonstrate a convincing performance and detect only about 10 percent of the potential accidents.

3.7.3 Discussion

The study of the effect of network density finds that when more vehicles enter into the same enclosed area, competition for accessing the medium gets stiffer and streets with more cars tend to have more collisions. This behavior suggests that as network density increases, the number of rebroadcasts of warning messages significantly increases in the network, ultimately resulting in a broadcast storm. These findings then relate collision with queuing delay and show that a collision is a cause generated by a broadcast storm and a queuing delay is its consequences. This study also demonstrates that although initially jitter increases sharply, it becomes stable as network density increases. As jitter is the measurement of the queuing delay's standard deviation, this behavior indicates with broadcast storm becoming severe delays tend to fluctuate less.

Therefore, based on the network layer observation, this study confirms that collisions occur more frequently as network density increases. This behavior subsequently results in considerable queuing delays in the network. It further summarizes that when the AWSs operate with 5 warnings/second data traffic load, they fail to comply with the two-second rule in high network densities. The systems were also evaluated with a heavy load of 15 warnings/sec to observe their performance under stressed environments. The outcome shows that they fail to comply with the two-second rule, even in low network densities; hence the first research question is addressed.

The observation made in the application layer finally sums up the chapter by showing that failing to comply with the two-second rule has severe consequences as AWSs cannot successfully detect potential accidents in the simulation. If the AWS cannot control the broadcast storm, it remains a risk that pushes the queuing delay towards, or more than, MTQD and subsequently makes the AWS vulnerable. Therefore, the AWS must introduce an improvement in the broadcast scheme to maintain transmission queuing delay below MTQD to comply with the two-second rule. With this finding, the chapter obtains the answer to the second research question.

3.8 CONCLUSION

Broadcast is the most apparent data dissemination method for AWSs but tends to generate broadcast storms in the network that can isolate nodes for some period. As the degree of this threat was unknown, it was essential to investigate how badly it might affect AWSs, particularly against the two-second rule. This chapter has accomplished that objective by showing that existing systems cannot handle excessive transmissions and often exhibit long transmission queuing delay while going through broadcast storms. This behavior leads to the system's collapse for more than the critical threshold of MTQD in several situations. During these phases, vehicles can move along road segments and pass through junctions with little or no communication ability with their neighbors. This chapter's findings demonstrate that broadcast schemes are not reliable enough to disseminate warnings in AWSs consistently and an enhancement on top of broadcast schemes is a necessity. In recent time, several new technologies such as blockchain,[36, 37] machine learning,[38] the Internet of Things,[39] and decentralized data storage[40, 41] have been used to improve this performance.

REFERENCES

[1] Saif Al-Sultan, Moath M. Al-Doori, Ali H. Al-Bayatti, and Hussien Zedan. A Comprehensive Survey on Vehicular Ad Hoc Network. *Journal of Network and Computer Applications*:1–13, 2013.

[2] Niaz Morshed Chowdhury, Lewis M. Mackenzie, and Colin Perkins. Requirement Analysis for Building Practical Accident Warning Systems based on Vehicular Ad-hoc Networks. In *Proceedings of the 11th IEEE/ IFIP Annual Conference on Wireless On-demand Network Systems and Services (WONS)*, pages 81–88, Obergurgl, Austria, April 2014. IEEE.

[3] T.A. Ramrekha and C. Politis. A Hybrid Adaptive Routing Protocol for Extreme Emergency Ad Hoc Communication. In *Proceedings of 19th International Conference on Computer Communications and Networks*, Zurich, Switzerland, August 2010.

[4] Arzoo Dahiya and R. Chauhan. A Comparative Study of MANET and VANET Environment. *Journal of Computing*, 2:87–92, 2010.

[5] S.Y. Wang. Predicting the Lifetime of Repairable Unicast Routing Paths in Vehicle-Formed Mobile Ad Hoc Networks on Highways. In *Proceedings of the 15th IEEE International Symposium on Personal, Indoor and Mobile Radio Communications*, Barcelona, Spain, September 2004. IEEE.

[6] Brad Williams and Tracy Camp. Comparison of Broadcasting Techniques for Mobile Ad hoc Network. In *Proceedings of the 3rd International Symposium on Mobile ad hoc Networking and Computing (Mobihoc)*, pages 194–205, Lausanne, Switzerland, June 2002. ACM.

[7] Sarah Omar Al-Humoud. *The Dynamic Counter-Based Broadcast for Mobile Ad hoc Networks*. PhD Thesis, University of Glasgow, Glasgow, United Kingdom, 2011.

[8] P. Wei and L. Xi-Cheng. On The Reduction of Broadcast Redundancy in Mobile Ad hoc Networks. In *Proceedings of the First Annual Workshop on Mobile and Ad Hoc Networking and Computing*, Boston, MA, 2000.

[9] T. Yu-Chee, N. Sze-Yao, and S. En-Yu. Adaptive Approaches to Relieving Broadcast Storms in a Wireless Multihop Mobile Ad hoc Network. *IEEE Transactions on Computers*, 52:545–557, 2003.

[10] Z. Hao and J. Zhong-Ping. Performance Analysis of Broadcasting Schemes in Mobile Ad hoc Networks. *IEEE Communications Letters*, 8:718–720, 2004.

[11] Sze-Yao Ni, Yu-Chee Tseng, Yuh-Shyan Chen, and Jang-Ping Sheu. The Broadcast Storm Problem in a Mobile Ad Hoc Network. In *Proceedings of ACM MOBICOM*, pages 151–162, Seattle, WA, 1999.

[12] Jong-Mu Choi, Jungmin So, and Young-Bae Ko. Numerical Analysis of IEEE 802.11 Broadcast Scheme in Multihop Wireless Ad Hoc Networks. *Proceedings of ICOIN*, 3391:1–10, 2005.

[13] Yu-Chee Tseng, Sze-Yao Ni, Yuh-Shyan Chen, and Jang-Ping Sheu. The Broadcast Storm Problem in a Mobile Ad Hoc Network. *Wireless Networks*, 8:153–167, 2002.

[14] N. Wisitpongphan, O. K. Tonguz, J. S. Parikh, P. Mudalige, F. Bai, and V. Sadekar. Broadcast Storm Mitigation Techniques in Vehicular Ad Hoc Networks. *IEEE Wireless Communications*, pages 84–94, 2007.

[15] Francisco J. Martinez, Chai-Keong Toh, Juan-Carlos Cano, and Carlos T. Calafate. A Street Broadcast Reduction Scheme (SBR) to Mitigate the Broadcast Storm Problem in VANETs. In *Proceedings of Wirelss Personal Communication*, 2011.

[16] DVSA. Highways Agency Warns Tailgaters that 'Only A Fool Breaks the 2-second Rule.' Technical report, Driver and Vehicle Standards Agency, May 2014.

[17] RSA-Ireland. Driving Safely in Traffic – The Two Second Rule. Technical report, Road Safety Authority, Republic of Ireland, 2014. URL: http://www.rotr.ie/rules-for-driving/.

[18] DMV-NY. Chapter 8: Defensive Driving. Technical report, The New York State Department of Motor Vehicles, 2015. URL: http://dmv.ny.gov/about-dmv/chapter-8-defensive-driving.

[19] Niaz Morshed Chowdhury. *NETCODE: An XOR-based Warning Dissemination Scheme for Vehicular Wireless Networks*. PhD thesis, University of Glasgow, Glasgow, United Kingdom, 2016.

[20] Francisco J. Martinez, Juan-Carlos Cano, Carlos T. Calafate, and Pietro Manzoni. A VANET Solution to Prevent Car Accident. In *Proceedings of Jornadas de Paralelismo*, Spain, 2007.

[21] Lin Yang, Jinhua Guo, and Ying Wu. Piggyback Cooperative Repetition for Reliable Broadcasting of Safety Messages in VANETs. In *Proceedings of 6th IEEE Consumer Communications and Networking Conference*, Las Vegas, NV, 2009.

[22] Dorothy Stanley. 802.11-2012 - IEEE Standard for Information Technology. Technical report, 2012. Standards Committee: C/LM - LAN/MAN Standards Committee.

[23] USC. The Network Simulator – ns-2. Technical report, The University of Southern California, November 2011.

[24] Riverbed. Riverbed Modeler: The fastest discrete event-simulation engine for analyzing and designing communication networks. Technical report, Riverbed Application and Network Performance Management Solutions, 2015.

[25] OpenSim. OMNeT++ 5.0b3 released. Technical report, OpenSim Ltd, December 2015.

[26] Shiwen Mao. Fundamentals of Communication Networks: Principles and Practice. In *Cognitive Radio Communications and Networks*, pages 201–234. Academic Press (AP), 2010. Section: Chapter 8 – Fundamentals of communication networks.

[27] C. Sommer, R. German, and F. Dressler. Bidirectionally Coupled Network and Road Traffic Simulation for Improved IVC Analysis. *IEEE Transactions on Mobile Computing*, 10(1): pp. 3–15, January 2011.

[28] S. Kurkowski, T. Camp, and M. Colagrosso. MANET Simulation Studies: The Incredibles. *ACM SIGMOBILE Mobile Computing and Communications Review*, 9(4), 2005.

[29] Mark Ciampa. *CWNA Guide to Wireless LANs*. Cengage Learning, 2013.

[30] F.J. Martinez, Chai-Keong Toh, J.-C. Cano, and C.T. Calafate. Realistic radio propagation models (RPMs) for VANET Simulations. In *Proceedings of the Wireless Communications and Networking Conference (WCNC)*, pages 1–6, Budapest, Hungary, April 2009.

[31] C. Sommer, D. Eckhoff, R. German, and F. Dressler. A computationally inexpensive empirical model of IEEE 802.11p radio shadowing in urban environments. In *Proceedings of the Eighth International Conference on Wireless On-Demand Network Systems and Services (WONS)*, pages 84–90, Bardonecchia, Italy, 2011.

[32] M. Boban, T.T.V. Vinhoza, M. Ferreira, and J. Barros. Impact of Vehicles as Obstacles in Vehicular Ad Hoc Networks. *IEEE Journal on Selected Areas in Communications*, 29(1):15–28, 2011.

[33] A. Mahajan, N. Potnis, K. Gopalan, and A. Wang. Modeling vanet deployment in urban settings. In *Proceedings of the 10th ACM Symposium on Modeling, Analysis, and Simulation of Wireless and Mobile Systems*, pages 551–558, Crete Island, Greece, 2007.

[34] T. Camp, J. Boleng and V. Davies. A survey of mobility models for ad hoc network research. *Wireless Communication and Mobile Computing Special Issue on Mobile Ad Hoc Networking: Research, Trends and Applications*:483–502, 2002.

[35] Mirco Musolesi and Cecilia Mascolo. Mobility Models for Systems Evaluation: A Survey. Technical report, Dartmouth College (USA) and University of Cambridge (UK), 2011.

[36] Niaz Chowdhury. *Inside Blockchain, Bitcoin, and Cryptocurrencies*. Taylor & Francis, 1st edition, 2019.

[37] Md Sadek Ferdous, Mohammad Jabed Morshed Chowdhury, Kamanashis Biswas, Niaz Chowdhury, and Val- lipuram Muthukumarasamy. Immutable Autobiography of Smart Cars Leveraging Blockchain Technology. Knowledge Engineering Review, 35 (3):1–17, January 2020. Cambridge University Press.

[38] Michael Slavik and Imad Mahgoub. Applying machine learning to the design of multi-hop broadcast protocols for VANET. In *Proceedings of the 7th International Wireless Communications and Mobile Computing Conference*. IEEE, July 2011.

[39] John Moore, Gerd Kortuem, Andrew Smith, Niaz Chowdhury, Jose Cavero, and Daniel Gooch. DevOps for the Urban IoT. In *Proceedings of the Second International Conference on IoT in Urban Space*, pages 78–81. ACM, May 2016.

[40] Manoharan Ramachandran, Niaz Chowdhury, Allan Third, John Domingue, Kevin Quick, and Michelle Bachler. Towards complete decentralised verification of data with confidentiality: Different ways to connect solid pods and blockchain. In *Proceedings of the Web Conference*, pages 645–649, Taipei, Taiwan, April 2020. ACM.

[41] Manoharan Ramachandran, Niaz Chowdhury, Allan Third, Zeeshan Jan, Chris Valentine, and John Domingue. A Framework for Handling Internet of Things Data with Confidentiality and Blockchain Support. Heraklion, Greece, May 2020.

Chapter 4

The uses of big data in smart city transportation to accelerate the business growth

Md Khaladun Nabi

Birmingham City University, Birmingham, United Kingdom

CONTENTS

4.1 INTRODUCTION

Big data is one of the most fascinating and exhilarating enhancements in the present transportation industry. As a matter of fact, in numerous ways, the future of the smart city transportation industry has already happened. The transportation industry is evolving towards big data analytics to find effective

DOI: 10.1201/9781315110905-4

and smarter ways to utilize resources that exist, decreasing overcrowding and increasing the travel experience. Sometimes living and roaming around in big metropolitan cities can be an actual dilemma because of too much traffic, overcrowding, and inefficient transportation facilities. While big data has existed in the business word for years, traditional approaches for handling transportation might be on the verge of change, thanks to big data advances. According to (1), the development of big data in smart city transportation will lead transportation industries to reconsider job descriptions and assess when humans should be at the wheel, and when the machines should drive, for safety, cost saving measures, and skills. For example, in the United States, logistic companies have already started working hard on using sensor data in vehicles to accession directing and decrease fuel consumption. The American company US Xpress installed approximately 1,000 data sensors to monitor trucks, such as the trucks' speed, breaks, maintenance, and driver capabilities (2). Using Big data, large logistic companies can now track hundreds of thousands of vehicles, enabling companies to keep an eye on trucks' time management, the company's inventory, and the trucks' destinations. This statistic helps transportation companies enhance their task force, increasing their capability to accelerate business growth. Big data analytics in smart city transportation have literally changed and fasten merchandise, logistic, and supply chain management businesses. The world has entered the digital age and the big data analytics are the latest technology to achieve great strides. It is predicted that by the end of 2020, big data volume is going to reach 44 trillion gigabytes, flouting previous trends and potentially creating a new world of business (3). The focus of this chapter is on ways big data can transform the smart city transportation system to expand business development and growth strategies. Additionally, this chapter explores using big data in a smart city transportation system to benefit business expansion and international growth.

4.1.1 Transform the smart city transportation system with big data for modern business development

Quality of life and business development is not great when people and vehicles cannot move effectively in smart cities. Disorganization costs money, increases pollution, and takes time away from people's lives. The transportation infrastructure develops more sluggishly than the demand for it, causing a slower growth of business communication and expansion. For this reason, transportation industries and some countries, as well as governments, are turning to big data analytics to identify the smartest ways to use the resources that exist, reduce traffic congestion and improving the travel experience as well as business development overall. Using big data in smart transportation systems can help cities develop sustainability, governance, an enriched quality of life, the intellectual management of natural resources, business development, and city services.

4.1.1.1 Increasing efficiency through big data in smart city transportation

Lake of parking, long exchanges, and overcrowding are central problems to city transportation systems that obstruct business productivity and progression. However, using big data in a smart city transportation system can pinpoint causes of the crowding, identify where the driver will park the vehicles, and sometimes provide an alternate route. Big data is projected to increase the efficiency of the smart city transportation system, enhancing business industries to move everything more quickly for everyone (4). It uses a machine-learning method that will enable vehicles to communicate with one another, thereby reducing travel time and increasing business growth. For example, the European Union scheme of Transforming Transport Industry describes ways to use big data analytics to improve efficiencies and generate business growth and environmental benefits (5). Numerous areas of the big data application are observed in the scheme. In this scheme, big data identifies routes using other ways between hubs like Amsterdam and Frankfurt. Big data was able to monitor delivery consistency and faithfulness to identify problems and remove flawed processes. And in the end, big data analyzed a truck's travel time, traffic information, airport operations, and data from routing applications. Big data can provide detailed routing and forecasting data, especially in heavy goods vehicle (HGV) traffic, an estimation of truck arrivals that decreases the buffer time and enables fleet management to become more efficient, diminishing the overall vehicle fleet and saving labor costs.

4.1.1.2 Big data can identify dangers in smart city transportation

For business development, big data introduces more combined models and vehicles in the transportation industry, which, subsequently, increases safety issues. According to (6), big data can obstruct *right-hook accidents*, where a vehicle indicates a right-hand cross turn while there is a cyclist or another vehicle to their right, in a blind spot. This can cause serious accidents, but by using big data, vehicles can be more intellectual and consistent. Phillips (7) suggests that, in a smart city transportation system, big data have a commercial implementation that can be used to monitor the safety and security of the driver. Big data can predict the risk of the driver's speeding, seat belt usage, probability of breaking vehicles laws, and much more. Big data can also gather data to predict potential crash areas, giving the drivers city insights maps to recover the structure with safety as the priority. Big data can create monitoring technology, which can identify previously implemented, targeted enhancements towards safety, security, and reliability of the smart city transport infrastructure. Big data can indicate faults, cybercrime, and terrorism, and it can reduce these threats by providing a smart intelligence system such as closed-circuit television (CCTV) (8).

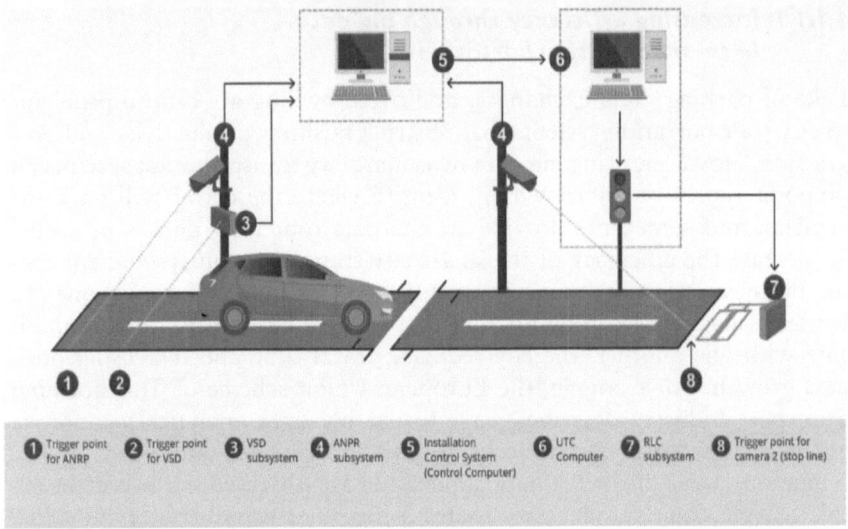

Figure 4.1 Security and danger inspection system in the smart city transportation.
(Source: [13])

Further, in smart city roads using big data technology in security cameras can easily manage properties and provide safety features such as detecting unusual occurrences and informing the police (see Figure 4.1). For example, a camera could identify somebody approaching a car with a weapon and intent to take the car. This data could rapidly be shared with the authorities, who could then start searching for the stolen vehicle (9).

4.1.1.3 Improving operational competencies with big data in smart city transportation

Big data in the smart city transportation industry impacts the overall operational efficiency, including warehousing, production, procurement, and the supply chain. Today, companies need to provide the cheapest rate of transportation to serve the customers while keeping the business profitable. To achieve this, sometimes it's very critical for businesses to improve their routes. To enhance the routing vehicles, crew, and goods, big data analytics are typically used to solve the warehouse solutions. Robins (10) suggests that sensor-driven data analytics provide more applicable evidence about the vehicles and the goods, which helps identify many more possible maintenance requirements and security issues. Sensors in the vehicles can predict real-time information about how the vehicles are performing, their speeds, and when they will need to be replaced (see Figure 4.2). With the big data analytics, sensors can evaluate the health of the engine and equipment, enabling businesses to identify problems and maintenance issues without losing too much time.

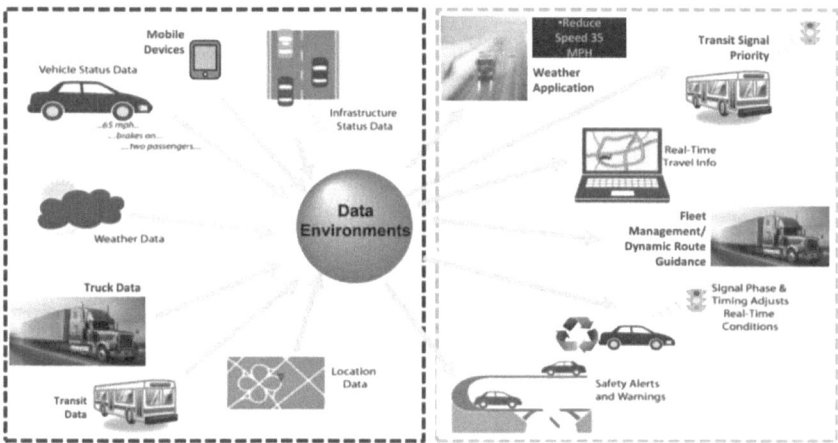

Figure 4.2 Big data in a smart city transportation operation.
(Source: [11])

4.1.1.4 Big data can improve e-commerce logistics

Big data can lead to new, innovative, and more advanced e-commerce, as well as logistic procedures that benefit organizations, online retailers, and customers. Using big data in the smart city transportation industry aids e-commerce organizations in their decision making, allows them to save on costs, and enhances their performance, products, and operating procedures. It can identify the delivery designs and predict future delivery demands, refine delivery details, and improve customer deliveries, including proposing alternative routes in the cities. Riddle (12) suggests that big data in smart city transportation helps companies in e-commerce businesses understand consumer buying behavior and optimize logistics and operating processes. Using big data in e-commerce helps businesses develop faster shipments, more personalized offerings, and better customer service. Big data can also predict fraudulent activities and shape product pricing.

4.1.1.5 Big data in autonomous vehicles in smart city transportation

By the end of 2021, it is projected that approximately 10 million self-driving cars will be on the streets in smart cities, and they will be automatically connected with high-tech networking systems and sharing road maps and other features Business Insider (13). Big data helps autonomous vehicles connect and report one dominant database system, where road specialists will identify the routes that save fuel, money, and make smart city transportation and deliveries much more efficient. Autonomous cars also provide ways to improve lifestyles, energy efficiency, environmental impacts, and customer

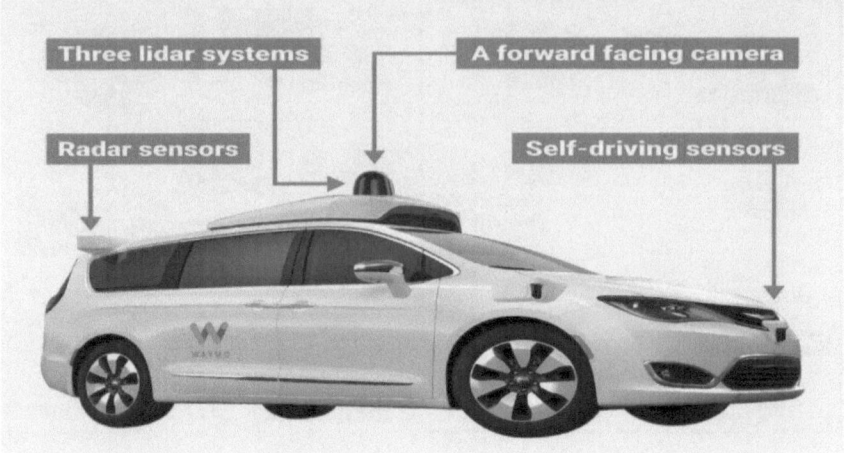

Figure 4.3 Autonomous vehicle.

(Source: Waymo.com)

safety (see Figure 4.3). Big data assists drivers, and autonomous cars task force managers to endorse effective driving and decrease CO_2 emissions. This technology also helps cars to run more effectively and smoothly by improving overcrowding in the smart cities and aids logistics, which increase business growth. Big data technologies like floating car data (FCD), dashboard-display software that can monitor and evaluate vehicles' task force actions, satellite images that forecast productivity, and enhanced evaluations of route times make the vehicles smarter and aid road transport (1). Additionally, Waymo (14) states that the website, Waymo.com has now registered the equivalency of 300 years of driving experience on the smart city streets and also promises time efficiencies, less congested roads, safer roads, and break-down recovery for the smart city.

4.1.1.6 Big data control traffic management system that increases business growth

No one likes traffic. Big data controls the total traffic management system and enhances traffic flow by forecasting and managing traffic jams. Big data can combine real-time information, historical data, and useful algorithms, as well as interpret vehicles speeds, weather situations, communal procedures, and road work activity. Parakh (15) states that traffic problems are one of the major infrastructure challenges tackled by developing countries today, whereas developed countries with smart cities are already using big data and Internet of Things (IoT) to minimize traffic issues (see Figure 4.4). Numerous countries are solving traffic congestion using attractive information from CCTV feeds and conveying vehicle-related statistics to city traffic management centers to assist with improvements.

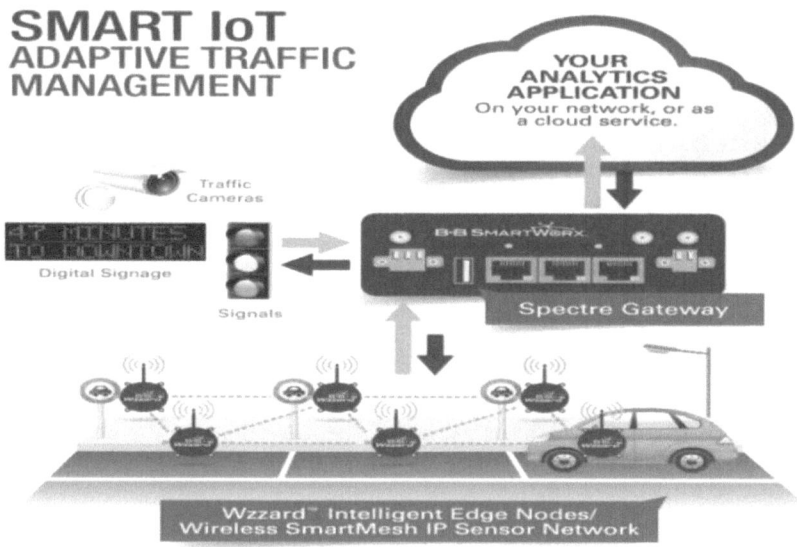

Figure 4.4 Use of big data and IoT in a traffic management system.
(Source: [16])

Better-organized traffic systems mean better movement of vehicles on the road; in other words, no cars, buses, and trucks in traffic jams impeding business operations and growth. All this eventually translates to lower run times, the proper utilization of natural resources (gas), and less pollution.

Big data can upgrade the situational awareness of the city traffic as well. This involves refining the identification of traffic jams using traffic cameras, distributing key traffic information to travelers over social media, and creating a parking registration system for shipping delivery in the city, which helps the business operate smoothly. According to (17), big data can be used for numerous applications:

- **Distant monitoring** and real-time traffic flow data.
- **Build complex models of a city's transportation system** to help city planners create knowledgeable, data-driven results for substructure development and speculation.
- **Influence artificial intelligence (AI) and cloud computing** to regulate traffic light designs animatedly and in real time.
- **Provide the building blocks for connecting vehicles** and intelligent traffic systems (ITS).

With respect to better logistics and business expansion in urban areas, big data can build superior traffic replication micro-models from real data to help city authorities create the decision-making procedure. It is an innovative design tool for logistics companies and finds methods to improved and comprehend the behavior of logistics vehicles in city centers.

4.2 BIG DATA FOR SMART TRANSPORTATION TO ACCELERATE THE LOGISTIC AND SUPPLY CHAIN MANAGEMENT

Big data in smart transportation goes well with the logistics business. It is projected that logistic companies will experience considerable expansion due to big data analytics on their operational accomplishments. Information collected from various sources should be turned into something creative to aid decision making. Logistics service workers gather numerous data, such as the scope and the weight of packages, their destination, the present location of consignments, and so on. This procedure has become simple and error proof. It reveals the purchase patterns of diverse customers, the newest developments in the marketplace, and cycles of conservation. According to (18), using big data in smart city transportation will be an important asset in the future of logistic business, and its advantages are being realized today. The overall logistic and supply chain will lower operating costs, will make and test new innovative business models, will develop and maintain a higher standard in consumer satisfaction, and will attain unsurpassed effectiveness. Figure 4.5 shows the true scope of the impact of big data in the smart city transportation, supply chain, and logistics industries.

At one time, logistics and supply chain management procedures used to fail due to incorrect, slow, and unorganized management process. Nowadays, using big data analytics, supplier networks experience improved data accuracy, transparency, and comprehension, which and leads to enhanced appropriate intelligence through supply chains and logistics. The impact and benefits of big data in smart city transportation is described in the sections that follow.

4.2.1 Real-time fleet tracking

Logistic companies can use big data analytics, real-time fleet tags, Global Positioning System (GPS) devices, bar codes, and many more things to track the vehicles in real time, as well as real-time traffic data, road network systems, and fleet data. According to (19), by using real-time data collection in a workplace vehicle tracking system, business owners can confirm that delivery employees reach their destinations as speedily and safely as possible. Through this technology, possible dangers and delays can be assessed, and alternative options can be communicated using mobiles or computers. This addresses inadequacies, and more customers can be helped. Real-time fleet tracking systems also optimize routes, planning and scheduling deliveries that help the logistic managers. This technology also predicts foul weather,

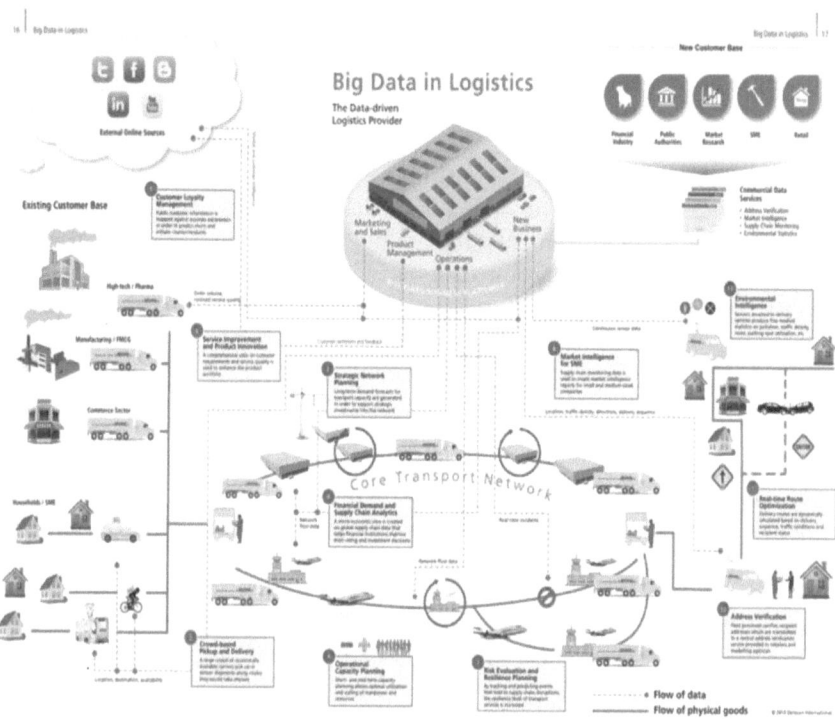

Figure 4.5 Big data in the transportation, supply chain, and logistics industries.
(Source: [18])

and tracks shipments and employees from anywhere. Further, Transfinder (20) specifies that GPS fleet tracking software does a lot more things than tracking the measure and the routes of the fleet; it also allows businesses to monitor, schedule, and assess drivers efficiently and monitor fuel usage, specifically when prices increase. These being said, fleet tracking software uses widespread data. In other words, you can get the same analysis using manual worksheets if the fleet is just one vehicle.

4.2.2 Warehouse management efficiency

Today, warehouse managing through Enterprise resource planning (ERP) system is outdated. It's twenty-first century and millions of customers are interested in getting real-time updates on products and orders, and knowing a product's accessibility before making their buying decision. Through big data, warehouse managers can recognize how customers' behavior changes and what prospects they often have from producers and supply-chain managers. Warehouse managers get a full view into their filling, receiving, and delivering processes. Using this information, managers can schedule distributions in ways that improve protection and reduce expenditures. Using big data in warehouse management not only leads to faster delivery times and enhances customer service, but it also helps an organization increase its profitability and a competitive advantage in the marketplace. Big data enables companies to perform tasks around the clock with high efficiency and competency. According to (21), using big data tools such as smart devices, sensors, and cameras in the warehouse, enables companies to track their asset movements, and using smartwatches, employee can get immediate notifications of operations, which helps them make informed decisions almost instantaneously. *Cobots*, IoT, can update warehouse managers on technical problems in real time as well.

4.2.3 Improved responsiveness

In smart city transportation, using big data analytics to improve responsiveness can directly impact the performance of supply chains and logistics management. A responsive supply chain confirms on-budget and on-time delivery of quality products to meet consumer demands. Big data analytics helps supply-chain managers comprehend the marketplace and the competition (see Figure 4.6). They have the opportunity to improve their response times to customer, gain better control over inventory, save money, and improve overall efficiency. According to (22), the performance of the supply chain is compressed through good responsiveness in supply chain management. Responsive supply chains ensure the on-time and budget-friendly distribution of high-quality products to meet customer demands. For example, Amazon uses big data in their supply chain to monitor and

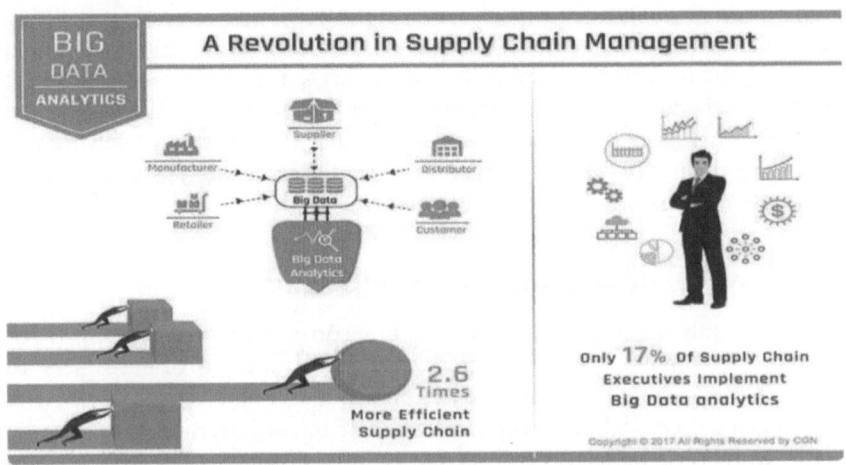

Figure 4.6 "Big data," a revolution in supply-chain management.

(Source: [23])

control an inventory of approximately 1.5 billion items in its 200 fulfill-ment centers. Wal-Mart, uses big data to handle millions of customer transactions per hour. It uses radio frequency identification (RFID) for tag-ging items and shipments. Consequently, the cohort of data is amplified to 100 to 1,000 times of data generated from straight bar code systems. It introduces data from various databases for advance analysis. In another example, UPS reformed their worldwide logistics network after the place-ment of *telematics* (a set of technologies used to analyze information from the vehicle) in its cargo segment (23).

The era of big data in the supply and logistics management industry in the smart city transportation system has just begun and there is a long way to go. Companies have to adopt the new technology sooner rather than later to conduct their supply chain experiments. Early adopters will experi-ence a modest control in connecting big data analytics to boost their supply chain performance.

4.2.4 Organized inventory management

Through big data analytics, warehouse managers and retailers get full insight information and product performance strategies. Forecasting the consumer demand for any particular product is possible. Big data helps managers evaluate the transparent supply chain and they can also track shipments. To reach the overall consumer demand, big data assists managers in avoid-ing overstocking by being aware of how much stock is needed. Inventory and supply chains are the strengths of any business, be it automobile, retail, manufacturing, pharma, and so on. Most industries have to stock some form of inventory and big data enables them to define when, what, and how much inventory they need to maintain the correct level. According to (24), big data includes the vast amount of data produced by countless sources, such as social media, mobile phones, dealings, and many more. By using big data in inventory management, related operations have also become more efficient than ever. Conversely, using data excessively without the appropriate tech-nology, substructure, and competency will not serve its purpose.

4.2.5 Seamless last-mile delivery

Due to mismanaged operations and rising costs, most logistic companies face challenges to progress with the last-mile delivery service. Using big data analytics in smart city transportation that examine the transfer to and contact with customers, logistics service providers can adjust their in-house procedures for last-mile deliveries better to make them fast and seamless. According to (25), these days a number of e-commerce companies have realized that they want to refine traditional last-mile delivery solutions in light of recent advances in data analytics and technology. Companies use big data from transactional records, delivery data from individual routes, high-resolution telemetry, and movement data from individual vehicles to create a more accurate account of their last-mile logistics operations. Using transportation networks through big data, companies are now able to obtain reliable, high-resolution, real-time visibility. Furthermore, this visibility is entirely accurate with regard to the position and location of shipments and the inventory levels of specific storage locations. According to (26), by 2030, inner cities will see a growing demand of 36 percent more delivery vehicles, which will lead to greater emissions and more traffic jams without effective intervention. Without effective intervention, city last-mile delivery releases and traffic congestion are on track to increase by more than 30 percent in the top 100 cities globally. The Future of the Last-Mile Ecosystem examines 24 improvements that can decrease emissions, congestion, and distribution costs for the urban last-mile. Through ecosystem-wide changes, interventions could diminish emissions and transportation congestion by 30 percent, and delivery costs by 25 percent, when compared with a "do-nothing" approach.

4.3 BIG DATA FOR SMART CITY TRANSPORTATION TO ACCOMPLISH THE EFFECTIVE CRM SCHEME

With a smart city transportation system, merchandise transporters and customers are given the information and tools to choose the best way to get their product from source to destination, through diverse modes of transport, with consideration to cost, time, and convenience. With the use of big data analytics in smart city transportation, customers can now control how the merchandise goes from A to B, and this will enable the customer to manage their supply chain as well as improve costs. It is really important to understanding the customer's needs and meet them to make a business successful. However, it is not always possible for companies to grow their businesses and keep up with every individual customer. For that reason, companies often find it problematic to retain their customers. However, with big data, the time has changed. In the logistics business, big data analytics benefit the companies through positive customer perception, a decrease in customer erosion, and an understanding about their demand. Using big data analytics in smart city transportation systems provides logistics companies with the right set of data about their customers. With this information in hand, companies can apply antique and predictive analytics models, and recognize how to generate customer devotion and improve the customer experience. As a result, they not only earn new customers, but also retain the current ones. According to (27), big data in customer relationship management (CRM) enables a company to enhance customer service, become more knowledgeable, calculate income on speculation, and forecast clientele attitude. The foremost objective of big data is to create a conglomerate between the interior CRM system and customer sentiment information that is external to the company's existing system. The CRM system also involves a company's interface with existing and future customers. Additionally, is also mixture of customer touchpoints. Using big data technologies, a CRM system can convert a true revenue driver. Rijmenam (28) states that big data can help CRM systems through managing the customer with organized data, such as an address, contact information, and newest contact; it also manages the customer's primarily inside-approach data by sending messages and storing basic information (see Figure 4.7). Conversely, big data assists the CRM system by interacting the customer using unstructured data such as e-mails, tweets, Facebook posts, comments, and so on. Analyzing the customers' activities with organized data, such as online visits, click-throughs, bound rates, and so on is an ambitious process that is mostly completed by analysts. Big data systems help analysts achieve more meaningful results because they provide data on a more regular, or real-time, basis.

Big data helps the company know their customer; where it really gets interesting is when big data engineers use a mixed data method to perform investigations, which allow the company to realize each customer individually on a real-time basis. They can deliver analytical approvals in order to progress/deliver the accurate product, for the accurate price, through the accurate channel. The outcome is an improved renovation rate. The use of big data in smart transportation can positively affect the company's CRM system. Big data foundations can come from organized or unstructured data arrangements. These data sources are collected from multiple channels, such as social networks, voice recordings, processed images, video recordings, open government data (OGD), and online customers' doings. According to Akter and Wamba (29), governments and private companies

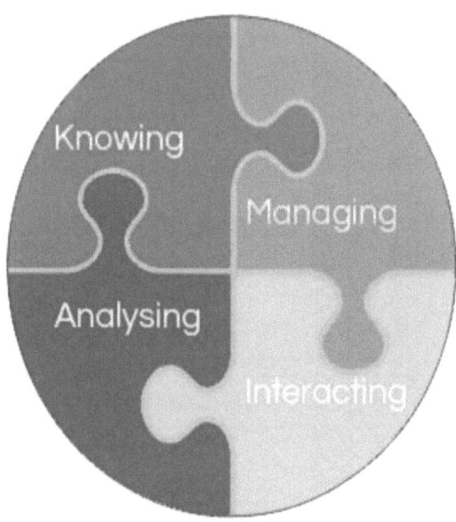

Figure 4.7 Big data and CRM.

(Source: [28])

see the potential of big data. Various organizations have made a huge investment to gather, integrate, and analyze data and use it to operate businesses. For this reason, in marketing and sales activities that are part of CRM's module, customers are visible, with a lot of marketing information each day, and although a lot of people just ignore the messages, companies still find a value in the messages received. Big data can support businesses in helping identify their customer base to advance their value, especially in sales, customer service, marketing, and promotion (30). The CRM system is a frontline in an organization that needs widespread, subsidiary, accurate data analytics to identify potential customers to engage with Sun, Strang, and Firmin (31). In this situation, big data's purpose is to support CRM policies, enabling organizations to quantify sales transactions, offer promotions, build product awareness, and build long-term customer relationships and loyalty. Given big data is able to provide customer profiles, businesses now can forecast and undertake the needs of their customers. Big data had helped form numerous industries and altered the way businesses operated today. Big companies such as Amazon definitely benefit from the shift, and technological giants like Google benefit from the volume of data they generate. Figure 4.8 shows how big data can contribute to a joint CRM strategy.

It is a new phenomenon for small and large businesses to collaborate with CRM and the big data influence has allowed for the availability and disposal of information. Big data deals with persistent information about achievement in CRM activities. Big data maintains the long-term connection by understanding customer preferences in a broad perspective. Customers can generate a huge collection of data related to their interests and preferences about products or services through numerous channels. Consequently, big data come up with an extraordinary view of customers so companies can

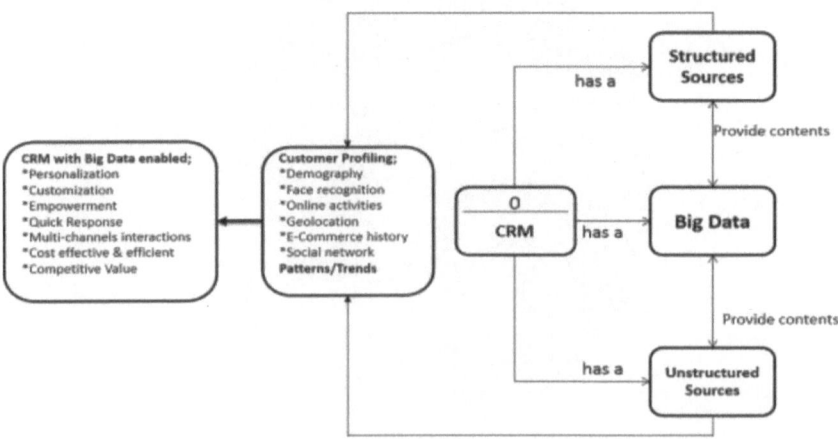

Figure 4.8 Basic framework of big data contributing to generate a CRM strategy.

(Source: [32])

increase customers' attention, satisfaction, engagement, participation, and personalization. Cleverism (33) specifies that CRM systems with big data analysis is an area of big transformation that can affect companies in delivering CRM strategies. There are numerous benefits to using big data in CRM systems, such as targeting customers, predicting how customers will react to marketing messages and product offerings, making personalized messages that create emotional attachments, and providing product offering, value chain, assessment measurement, digital marketing, and campaign-based strategies. According to (34), Netflix's successful usage of big data in CRM to run its streaming video service, instead of using outdated methods of data sharing, makes it capable for them to find out what they want, and made calculated marketing decisions.

4.4 CONCLUSION

The impact of big data in smart city transportation industries can define a new option and path to achieve business success. Big data in the transportation industry is on the brink of making the impossible possible. The most important improvements are likely to center on customer service as all of the advantages of big data point to a single, pleasant, and mutually-beneficial affiliation towards customers. In an organization, every individual employee involved in supply chain and logistic is employed to make the customer happy; and if the organization uses big data in all operations, the customers show loyalty and continue to purchase items from the company. Basically, big data is the greatest resource available to new supply chain entities, and as a result, big data should be used more to advance business-to-business and business-to-consumer relationships.

REFERENCES

1. Marr, B., 2020. *The Future of the Transport Industry—IoT, Big Data, AI and Autonomous Vehicles.* [online] Bernard Marr. Available at: <https://www.bernardmarr.com/default.asp?contentID=1204> [Accessed 7 December 2020].
2. Rijmenam, D., 2013. *Trucking Company US Xpress Drives Efficiency with Big Data.* [online] Datafloq.com. Available at: <https://datafloq.com/read/trucking-company-xpress-drives-efficiency-big-data/513> [Accessed 9 December 2020].
3. Anand, D., 2018. *How Big Data Science and Analytics Is the Lure for Businesses Today.* [online] Entrepreneur. Available at: <https://www.entrepreneur.com/article/316057> [Accessed 9 December 2020].
4. Kasturi, T., 2017. *Intelligent Trends for The Transportation and Industrial Enterprise—Insidebigdata.* [online] insideBIGDATA. Available at: <https://insidebigdata.com/2017/03/07/intelligent-trends-transportation-industrial-enterprise/> [Accessed 11 December 2020].

5. Gust-Kazakos, P., 2019. *Big Data in Transport Logistics | Transforming transport | PTV Blog*. [online] PTV Blog. Available at: <https://blog.ptvgroup.com/en/transport-logistics/gaining-time-big-data-transformingtransport/> [Accessed 14 December 2020].

6. Abels, D., 2018. *The Dangers of Right Hook Accidents — Chicago Injury Blog — October 3, 2018*. [online] Abels & Annes, P.C. Available at: <https://www.daveabels.com/the-dangers-of-right-hook-accidents/> [Accessed 12 December 2020].

7. Phillips, A., 2018. *How Big Data Is Helping Drivers Stay Safer On the Road—Insidebigdata*. [online] insideBIGDATA. Available at: <https://insidebigdata.com/2018/12/27/big-data-helping-drivers-stay-safer-road/> [Accessed 13 December 2020].

8. Rayner, T. and Bonas, W., 2017. *Deriving Transport Benefits from Big Data and The Internet of Things in Smart Cities | Lexology*. [online] Lexology.com. Available at: <https://www.lexology.com/library/detail.aspx?g=20f0bcc7-e9ea-4fad-9c00-4dd69a1b363d> [Accessed 14 December 2020].

9. Soares, E., 2019. *Intelligent Mobility: How Technology Is Solving Challenges for Smart Cities—Intellias*. [online] Intellias. Available at: <https://www.intellias.com/intelligent-mobility-how-technology-is-solving-challenges-for-smart-cities/> [Accessed 26 January 2021].

10. Robins, C., 2019. *The Most Innovative Executives of Trucking Industry 2019*. [online] Robins Consulting. Available at: <https://www.robinsconsulting.com/the-most-innovative-trucking-and-logistics-executives-of-2019/> [Accessed 14 December 2020].

11. Excellence, N., 2021. *Big Data and TSM&O | National Operations Center of Excellence*. [online] Transportationops.org. Available at: <https://transportationops.org/BigData/BigData-overview> [Accessed 26 January 2021].

12. Riddle, J., 2020. *How Will Big Data Transform E-Commerce Marketplaces?* [online] Learn.g2.com. Available at: <https://learn.g2.com/big-data-ecommerce> [Accessed 4 January 2021].

13. Business Insider. 2016. *10 Million Self-Driving Cars Will Be On The Road By 2020*. [online] Available at: <https://www.businessinsider.com/report-10-million-self-driving-cars-will-be-on-the-road-by-2020-2015-5-6> [Accessed 19 January 2021].

14. Waymo. 2021. Home – Waymo. [online] Available at: <https://waymo.com/> [Accessed 19 January 2021].

15. Parakh, M., 2018. *How IOT and Big Data Are Driving Smart Traffic Management and Smart Cities—DZone IoT*. [online] dzone.com. Available at: <https://dzone.com/articles/how-iot-and-big-data-are-driving-smart-traffic-man> [Accessed 27 January 2021].

16. Advantech IIoT Blog. 2017. *Smart IoT Technologies for Adaptive Traffic Management Using a Wireless Mesh Sensor Network—Advantech IIoT Blog*. [online] Available at: <https://blog.advantech.com/sites/iiot-us/smart-iot-technologies-for-adaptive-traffic-management-using-a-wireless-mesh-sensor-network/> [Accessed 27 January 2021].

17. Shell, C., 2019. Digitizing Traffic Management for Smart Cities – From Moving Vehicles to Moving People | Cleantech Group. [online] Cleantech.com. Available at: <https://www.cleantech.com/digitizing-traffic-management-for-smart-cities-from-moving-vehicles-to-moving-people/> [Accessed 27 January 2021].

18. Jessop, K., 2020. *The Impact of Big Data in the Transportation & Supply Chain Industries*. [online] Transportation Management Company | Cerasis. Available at: <https://cerasis.com/big-data-in-the-transportation/> [Accessed 30 January 2021].

19. Trackyourtruck.com. 2021. *Company Vehicle and Car Tracking System | Track Your Truck*. [online] Available at: <https://www.trackyourtruck.com/solutions/car-tracking/> [Accessed 30 January 2021].

20. Transfinder.com. 2021. *School Bus Routing Software | Transportation GPS | Transfinder from Transfinder*. [online] Available at: <https://www.transfinder.com/> [Accessed 30 January 2021].

21. Joshi, N., 2019. *The role of IoT in making warehouses | building a smarter warehouse |*. [online] Allerin.com. Available at: <https://www.allerin.com/blog/exploring-the-role-of-iot-in-making-warehouses-smarter> [Accessed 31 January 2021].

22. Deliforce.io. 2020. *Impact of Big Data on Logistics & Supply Chain Management | Deliforce*. [online] Available at: <https://www.deliforce.io/blog/impact-of-big-data-on-logistics-supply-chainmanagement#:~:text=Improved%20responsiveness&text=Responsive%20supply%20chains%20make%20sure, their%20companies%20and%20market%20scenario> [Accessed 31 January 2021].

23. Team, C., 2017. *Big Data Analytics: A Revolution in Supply Chain Management*. [online] Cgnglobal.com. Available at: <https://www.cgnglobal.com/india/the-edge-factor/big-data-analytics-a-revolution-in-supply-chain-management> [Accessed 31 January 2021].

24. Agarwal, Kriti. 2018. *How big is Big Data in Inventory Management—Orderhive*. [online] Available at: <https://www.orderhive.com/how-big-is-big-data-in-inventory-management> [Accessed 31 January 2021].

25. Winkenbach, M., 2018. *The Data Analytics Revolution in LAST-MILE Delivery—Supply Chain 24/7*. [online] Supplychain247.com. Available at:<https://www.supplychain247.com/article/the_data_analytics_revolution_in_last_mile_delivery> [Accessed 1 February 2021].

26. Chain 24/7, S., 2021. *The Future of the Last-Mile Ecosystem—Supply Chain 24/7 Paper*. [online] Supplychain247.com. Available at: <https://www.supplychain247.com/paper/the_future_of_the_last_mile_ecosystem> [Accessed 1 February 2021].

27. Contributor, T., 2015. *What is big data CRM (big data customer relationship management)? Definition from WhatIs.com*. [online] SearchCustomer Experience. Available at: <https://searchcustomerexperience.techtarget.com/definition/big-data-CRM-big-data-customer-relationship-management> [Accessed 5 February 2021].

28. Rijmenam, D., 2019. *How Big Data Turns CRM into Something Truly Valuable*. [online] Medium. Available at: <https://medium.com/dataseries/how-big-data-turns-crm-into-something-truly-valuable-c44f55482557> [Accessed 5 February 2021].

29. Akter, S. & Fosso Wamba, S., 2016. Big data analytics in e-commerce: A systematic review and agenda for future research. *Electronic Markets*, 26: 173–194.

30. Orenga-Roglá, S., Chalmeta, R., 2016. Social customer relationship management: taking advantage of Web 2.0 and Big Data technologies. *SpringerPlus*, 5: 1462.

31. Sun, Z., Strang, K. and Firmin, S., 2016. Business Analytics-Based Enterprise Information Systems. *Journal of Computer Information Systems*, 57(2): 169–178.
32. Anshari, M., Almunawar, M., Lim, S. and Al-Mudimigh, A., 2019. Customer relationship management and big data enabled: Personalization & customization of services. *Applied Computing and Informatics*, 15(2): 94–101.
33. Cleverism. 2021. Best Uses of Big Data in Marketing. [online] Available at: <https://www.cleverism.com/best-uses-big-data-marketing/> [Accessed 6 February 2021].
34. SchectmanReporter, J., 2012. *Netflix Uses Big Data to Improve Streaming Video*. [online] WSJ. Available at: <http://blogs.wsj.com/cio/2012/10/26/netflix-uses-big-data-to-improve-streaming-video> [Accessed 6 February 2021].

Chapter 5

A genetic blockchain approach for securing smart vehicles in quantum era

Bannishikha Banerjee, Ashish Jani, and Niraj Shah

School of Engineering, PP Savani University, Surat, India

CONTENTS

5.1 INTRODUCTION

Adroit automotive industry innovation and intellectualization is quickly emerging within urban populations. Advances in the realm of substantial intellectualization of the vast vehicular ecosystem and the (Huang et al., 2018 [1]) transport industry with the emergence of the smart vehicles has stirred up the brilliant minds of scientists (Goryaev et al., 2018 [2]) for developing smart vehicles and making them independent of human mistakes and more efficient when choosing routes. This will work perfectly (Gorodokin et al., 2017 [3]) with the proposed smart vehicles that are driven by the automated onboard processors and not individuals. It can be implemented by considering the potential gains of independent vehicles (Chłopek et al., 2014 [4]), such as providing more sophisticated alert management (Skrúcaný et al., 2019 [5]), smart sensors, and restricted intervals. Vehicle routes would become unambiguous and risk free (Skrúcaný et al., 2019 [5]) given the vehicles would not break any traffic norms. The breakthrough in the smart vehicle field (Skrúcaný et al., 2019 [5]) would be to associate them in a well-connected network (Bukova et al., 2018 [6]) of these vehicles, for improved coordination and information sharing. The neoteric furtherance in the locale of sensible conveyance and smart machinery (Narbayeva et al. 2020 [7]) stipulates profuse strategies for the authenticated relaying of vehicle-related information (Narbayeva et al. 2020 [7]). Connecting the sensible vehicles in a blockchain framework (Narbayeva et al. 2020 [7]) is sufficient to secularize the conveyance location and commuter-related information until the arrival of quantum computers (Gao et al. 2018 [8]).

DOI: 10.1201/9781315110905-5

Almost all of the modernized vehicles have onboard processors and inbuilt IoT devices that can be used (Jiang et al. 2017 [9]) to execute the blockchain architecture for the intellectualization of the conveyances without the need of any auxiliary equipment (Rathee et al. 2019 [10]). Blockchain connectivity would intellectualize the vehicles, by sharing the road images and location of the vehicles on the blockchain, hence reducing accidents. Therefore, the strategies for confidentiality preservation of the vehicular data should be prepared (Stopka, 2019 [11]).

Furthermore, the execution of blockchain courses of action is entirely wise and does not require extra hardware (Stopka et al., 2018 [12]). But intellectualization of conveyances (Madhushan et al., 2020 [13]) is directly proportional to the reliance on the internet, and, hence, the security issues. Therefore, cryptic approaches should be kept in mind, before we increase our dependency on smart vehicles. Current cryptographic architectures are functional only until the materialization of the much-anticipated quantum computers. In this work, we have developed a genetic approach-based post-quantum cryptic advancement that uses the Fourier's series for the mutation of the plaintext and binomial expression for the crossover of the resulting mutated data to be shared on the public blockchain among the smart vehicles.

Blockchain pledges the thwarting of the refashioning of data without relying on any third-party trusted centralized arbitrator, whereas the cloud does not assure complete integrity and tamper-free data. Cloud computing can jostle the execution of the blockchain framework-based administrations.

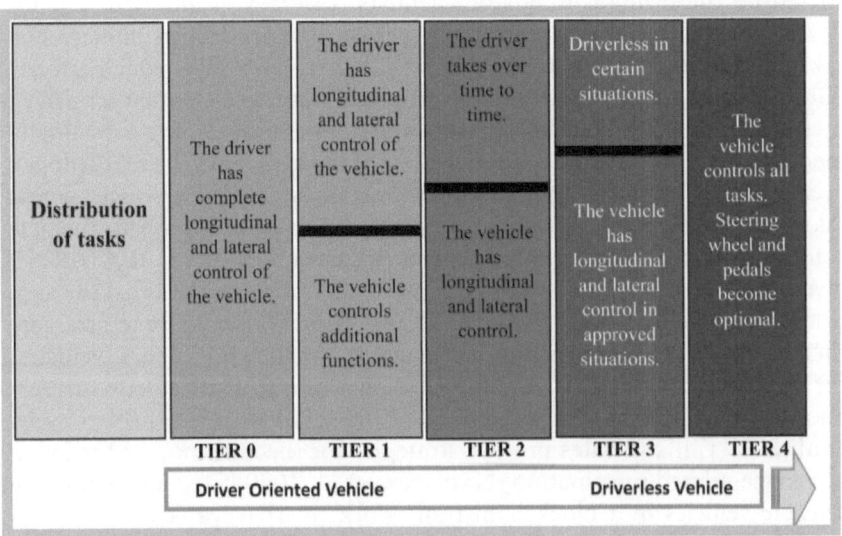

Figure 5.1 Tiers of self-driving vehicles.

Nonetheless, a cloud-based infrastructure has an amalgamate technique of data amassing, as the entire data prevails in a camaraderie's centralized set of data centers. Furthermore, the blockchain substructure has a quintessential philosophy of decentralization that connotes it does not store any of its information in one space (Park and Park, 2017 [14]). When considering hop-by-hop conduct, blocks of data are readdressed from junction to junction in a store-and-forward fashion. However, hop-by-hop transport has an indispensable desideratum of per-flow state information at intermediate nodes, which curbs its scalability. Furthermore, this emanates to the caching of factoids, which often ensures the ever-escalating vulnerability of the stored information. Henceforth, we can culminate that employing a blockchain is a befitting resort to the consolidated cloud and hop-by-hop system for secularization of the smart transport industry. Figure 5.2 and Table 5.1 demonstrate the difference between a cloud and blockchain.

In presence of quantum processors, the illustrious part of the already prevailing estimations that were sanguinely assumed to be secure until now may sputter, the elucidation being quantum figuring techniques; in other words, the brilliant Grover's elucidation and anticipated Shor's approach that bestow momentum to the cycle of calamitous constraining any computation. Genetic computations are stochastic explorations of Darwin's legendary postulation that fortuitous the enumeration on the embryonic orchestrations of facets and are instituted on the mechanics of populace genetic streak.

Interfacing the smart automobile's inbuilt IoT apparatus in a private blockchain arrangement and utilizing a genetic regimen to accord security covenants so that the best arrangement is acknowledged. Given hereditary methodology chooses the best answer for the transformation and hybrid of pieces, and it is incredibly troublesome or somewhat difficult to savage power, the most ideal arrangement is subsequently the mingling of a blockchain and hereditary methodology transpire to be sensibly difficult to break utilizing quantum processing assault. In this chapter, smart conveyance

Figure 5.2 Cloud versus blockchain.

Table 5.1 Blockchain vs. cloud Vs. hop-by-hop system [37]

Properties	Blockchain	Cloud	Hop-by-hop system
Immutability	Supported	Not supported	Not supported
Decentralized	Supported	Centralized authority	Centralized authority
Enhanced security	Hashing is used to provide authentication	Not supported	Not supported
Distributed ledgers	Leads to no malicious changes and verified ownership	Not supported	Not supported
Tamper free	Prevents tampering of data without relying on third party	Not supported	Not supported
Transparency	Supported	Not supported	Not supported
Consensus	Helps in decision making process	Not supported	Not supported
Smart contract	Faster settlement	Not supported	Not supported
Vulnerable to quantum computing attack	Not vulnerable	Vulnerable	Vulnerable

is simulated in a constricted blockchain substructure and employed in a genetic approach to provide security to the vehicular data.

5.2 TRANSPORT INTELLECTUALIZATION

As indicated by the cognoscenti (Makarova et al., 2017 [14]), 250 million linked conveyances will soon show up on the streets around the world. Given every one of them will be outfitted (Makarova et al., 2017 [14]) with more than 200 keen sensors, an onboard PC, and a cloud-based database, the requirement for information (Tsybunov et al., 2018 [15]) such as for the executives will be as significant as maneuvering during times of heavy traffic (Tsybunov et al., 2018 [15]). The amalgamation of a blockchain infrastructure and the intellectual vehicular industry provides answers for the most clutching issues, particularly those identified with unwavering quality and security (Tsybunov et al., 2018 [15]) of communication. The extent of blockchain innovation will invigorate inventive arrangements (Makarova et al., 2017 [14]) throughout the car ecological system. The tighter the vehicle is connected in a nexus (Tsybunov et al., 2018 [15]), the more defenseless it is to possibly destructive digital assaults. The blockchain would have the befitting course of action to embellish information streams from programmers with the most elevated accessible degree of security. In the vehicular innovation pursuit, the utilization of sequestrated records can guarantee that counterfeit scrap cannot be embedded into the pliant genuine fraction, which is undertaken by straightforwardness in the suave chain for creation, conveyance, and providers. Moreover, in the vehicle-blockchain combo business, perspicacious agreements can be implanted underway the blockchain to consequently give information requests at specific phases of the creation cycle. Transport industry intellectualization calls for real-time traffic and travel data (Khasanov et al., 2019 [16]), and constant data contributes altogether to the development of populace portability. The specifics are progressively maneuvered by satellite route administrations (Khasanov et al., 2019 [16]) and are presently offered as open and private sources. Given the rule of participation between moving members and components of foundation, including frameworks that give the correlation and dealing of particulars between vehicle-to-vehicle, between vehicle-to-cloud (Ipakchi and Albuyeh, 2009) [17]), and between various components of the cloud framework, these stratagems (Ipakchi and Albuyeh, 2009 [17]) are being augmented by the Global Navigation Satellite System (Khasanov et al., 2019 [16]) for situating and timing.

To administer the potentiality for which the telematics frameworks are proposed (Ipakchi and Albuyeh, 2009 [17]), the information sent and retrieved by them must be kept secure. Incorporating a blockchain architecture is inescapable in this context, given it is necessary for assurances against programmer assaults. The nonpareil advances in the realm of blockchain is the popularization of the idea of a smart vehicle. The resourcefulness for sustentation of the information of the related high-tech conveyances is still under continuation. The principal tests utilizing locally available smart frameworks (Farhangi, 2010 [18]) previously demonstrated that it plunges the adversity (Khasanov et al., 2019 [16]) by over one-third, and halves the number of deadly accidents.

The relationship between driver assistance frameworks to the expansion of semi-independent and pre-programmed vehicles (Khasanov et al., 2019 [16]) is a pervasive mold and is elucidated by the yearning of designers to guarantee the steadiness and security of the vehicle framework. Simultaneously, the rise of new, specialized, innovative arrangements is analogous with the blossoming (Farhangi, 2010 [18]) of modernized techniques, the disposition of which may require new strategies and instruments. The quintessentially unorthodox smart conveyances (Ipakchi and Albuyeh, 2009 [17]), with a prevailing cognizance, are of uttermost priority. Nevertheless, (Danzi et al., 2019 [22]) can modish interrelated frameworks that assemble the domain of secularization (Danzi et al., 2019 [22]) due to association (Ipakchi and Albuyeh, 2009 [17]) with other sensible conveyances. It is essential to recognize potential dangers (Ipakchi and Albuyeh, 2009 [17]), to foresee the likelihood of their event, and to decide the potential outcomes. The postulations to impede the menacing occurrences (Danzi et al., 2019 [22]) and dwindling the solemnity of the reverberation in the conveyances (Danzi et al. 2019 [22]). Owing to these circumstances, there subsists an expanding desideratum to seek orchestration (Kang et al., 2018 [20]) in the field of network secularization that would dispose of or, if nothing else, curb the feasibility of unessential impedance in the control arrangement of both the vehicle and the framework overall. Contemporary supplementary telematics administrations can proffer indemnification representatives data (Singh et al., 2015 [24]) on the motorist's sphere, such as the interval of the expedition, acceleration and slowing methodologies, vehicular momentum, cornering conduct, and other data. The blockchain bestows a congealed method for gathering this information and consigning it in a protected, unaltered state. The pertinent conglomerate encounters a dexterous revamping of (Danzi et al., 2019 [22]) a cognate conveyance. As a spearhead in this discipline, syndicates are utilizing blockchain concomitant upheavals throughout the merchandising for all the ascendancy it offers (Danzi et al., 2019 [22]). (Biswas et al., 2014 [25]) have proposed a genetic approach for securing devices connected in a network. (Chowdhury et al., 2014 [26]) suggested advances in the area of accident prevention in the vehicles associated with an ad-hoc network. The blockchain revolution would hypothetically enfold breakthrough refinements apropos of the brilliant framework. Decentralized advances have consistently been a reason for some keen matrix advances (Ipakchi and Albuyeh, 2009 [17]). The combination of environmentally friendly power sources, energy stockpiling gadgets, and evolved vehicles into the electric-powered vectors has established an uninhibited exploration belt on control schedulers to address the aforementioned issues (Farhangi, 2010 [18]). The idiosyncratic and alluring proclivity of blockchain recasting fabricated impressive interest in investigating and embracing this innovation in shrewd networks (Dong et al., 2018 [19]) (see Figure 5.3).

5.3 SMART VEHICLES BLOCKCHAIN FRAMEWORK

Securitization of smart vehicles is done by incorporating the rental cars in the blockchain framework and performing a genetic approach on the same. Figure 5.4 demonstrates the connectivity that is proposed in this chapter.

The genetic approach is implemented by performing mutation and crossover by flipping and repositioning bits in the given data. To do the aforementioned, start by first splitting the data into n dimension bits.

$$x = 2^i - i$$

Where i ranges from 0 to n.

a_0 is the first bit of the data, a_1 is the second bit, and so on, till a_n.

b_0 is the first bit of the public key, b_1 is the second bit, and so on, till b_n.

L is the number of bits in the public key.

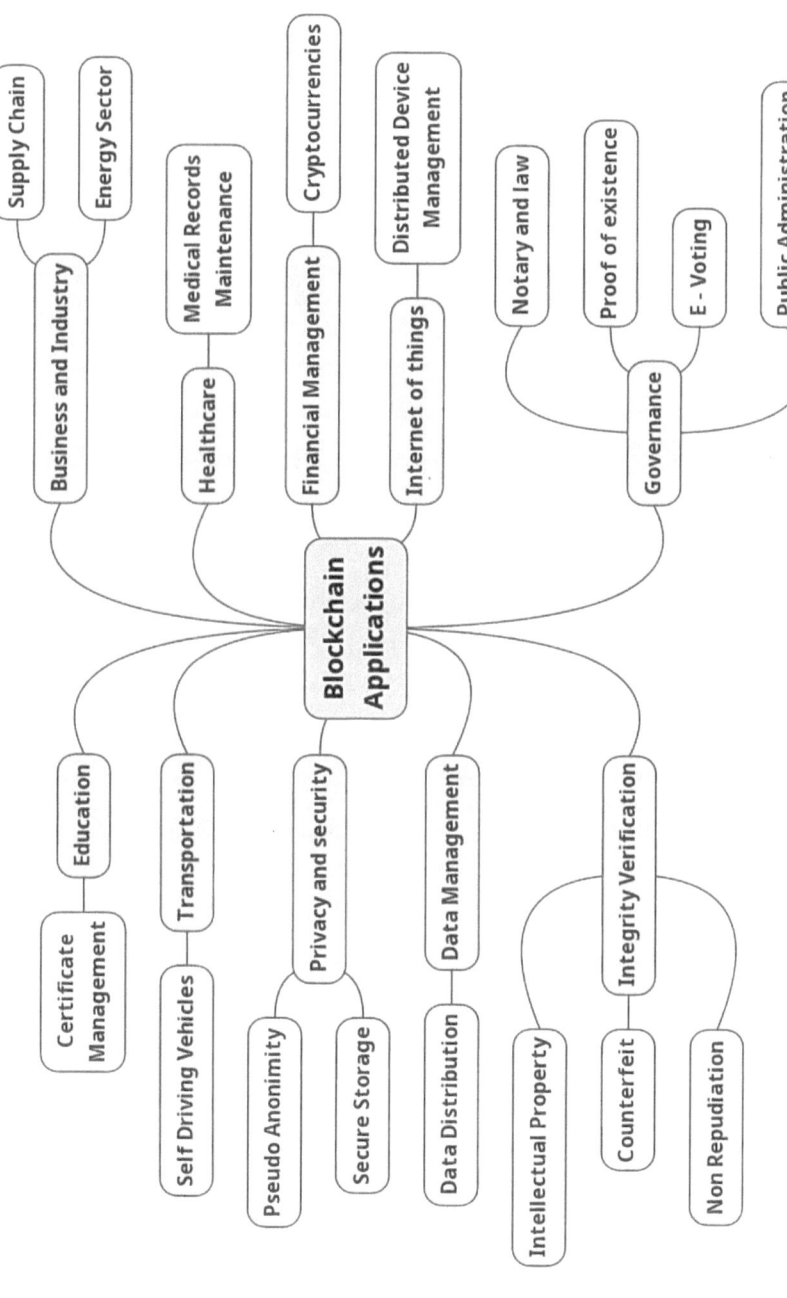

Figure 5.3 Mind map of blockchain applications.

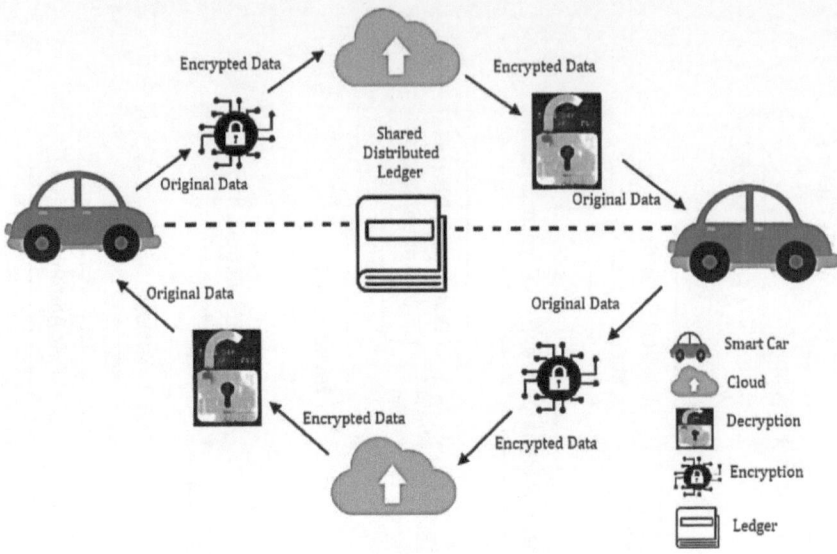

Figure 5.4 Smart car incorporated blockchain framework.

$$K = a_0 + \sum_{n=1}^{\infty}\left(a_n \cos\frac{n\pi x}{L} + b_n \sin\frac{n\pi x}{L} \right)$$

The resulting K is converted to a bitstream. Starting from a_1 and continuing with alternate bits, 0s are padded to reach n bits. The 1s in the bitstream determines the location in which bit flipping is done, which leads to mutation. The K is used to perform crossover. The following binomial equation is used for crossover (Table 5.2). a_0, and so on is taken from the mutated bitstream. L is converted to the bitstream. This bitstream is broken 2 *bits* streams.

$$L = (x + a)^n = \sum_{k=0}^{n}\binom{n}{k}x^k a^{n-k}$$

Table 5.2 Crossover

	Crossover			
Existing bits	00	01	10	11
Crossovered bits	10	11	00	01

If the bits are 00, then the crossover bit will be 10. If the bits are 01, then the crossover bit will be 11. If the bits are 10 then the crossover bit will be 00. If the bits are 11 then the crossover bit will be 01. Figure 5.5 shows the steps being followed for the computation of K and L. Furthermore, it shows mutation and crossover of the road image and vehicle location for ensuring security.

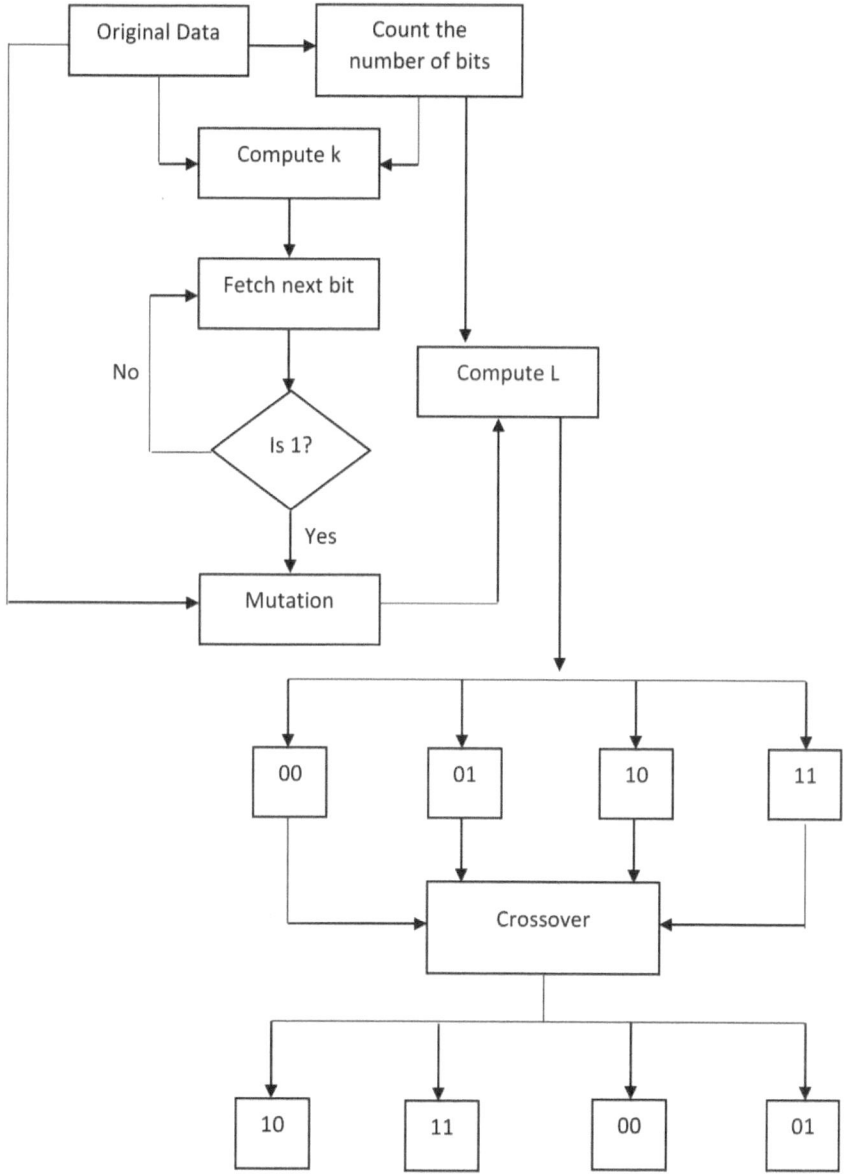

Figure 5.5 Flowchart of mutation and crossover.

Figure 5.6 Center image.

Conduction of this demonstrative investigation is done using Udacity's open-source smart vehicle simulation framework [34]. The model training is fulfilled scrupulously using the nano degree program for 10 minutes of drive, 10 times leading to an aggregate of 100 minutes total drive time. The car is outfitted with three additional homogeneous camcorders on its exterior to snap lane images from three different angles. Figures 5.6, 5.7, and 5.8 show the lane images snapped by the camcorders.

The screencaps are acessed using Open Source Computer Vision Library and converted to grayscale to diminish computational complexities (see Figure 5.9), as there are only two possible pixel values: 0 and 255.

Figure 5.7 Left image.

Figure 5.8 Right image.

After converting the image to grayscale, the next step is canny (see Figure 5.10). Canny differentiates the borders of the image. Any sharp change in intensity will be highlighted. This image is displayed using OpenCV.

We used matplotlib to display the canny image (see Figure 5.11) so that we can get the coordinates of the points on which the lanes will be drawn.

Figure 5.9 Grayscale image.

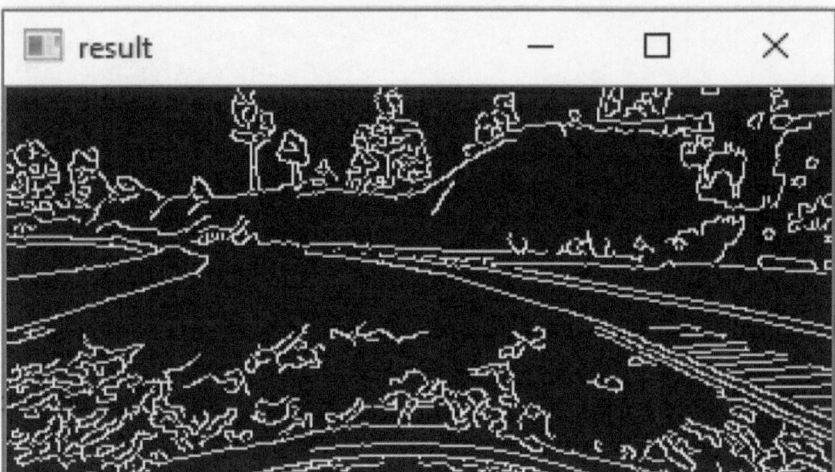

Figure 5.10 Canny image.

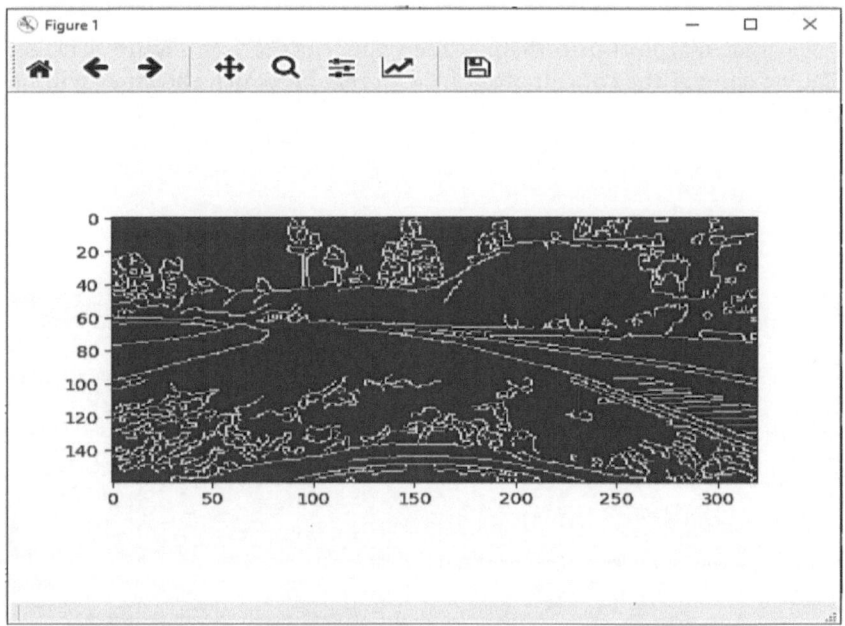

Figure 5.11 Canny using matplotlib.

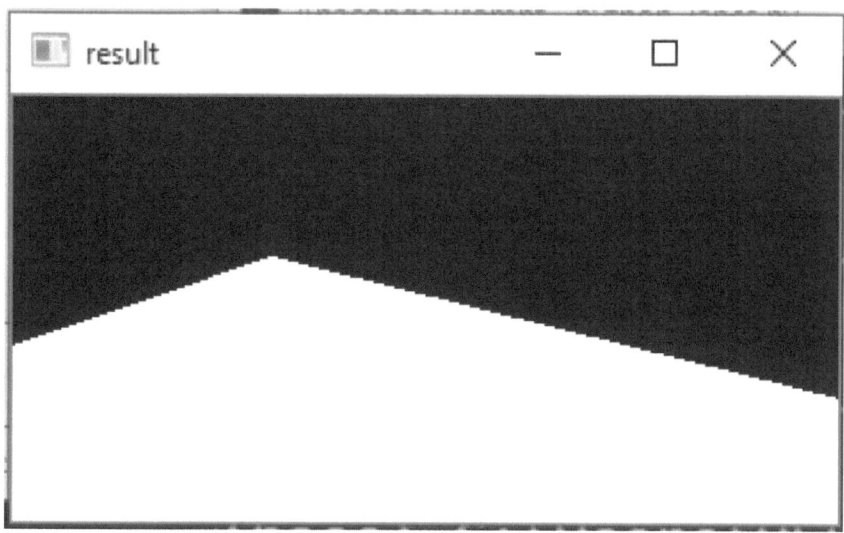

Figure 5.12 Region of interest.

Now we try to determine the region of interest by ignoring all the excess details and focusing only on the lane (see Figure 5.12).

$$laneshape = np.array\Big(\big[\big[(-200, height), (520, height), (100, 60)\big]\big]\Big)$$

$$mask = np.zeroes\_like(image)$$

For masking the AND operation is done along the coordinates. The points in the region of interest are given 1; the rest are 0. This results in 1s in the possible lane line region, ignoring all the lines that have lengths less than 1 or the distance between them is more than 5 (see Figure 5.13).

$$masked = cv2.bitwise\_and(laneshape, mask)$$

$$lines = cv2.HoughLines\Big(cropped_{image}, 5, np.\frac{pi}{180}, 100, np.array(\text{[]}),$$

$$minLineLength = 1, \max LineGap = 5\Big)$$

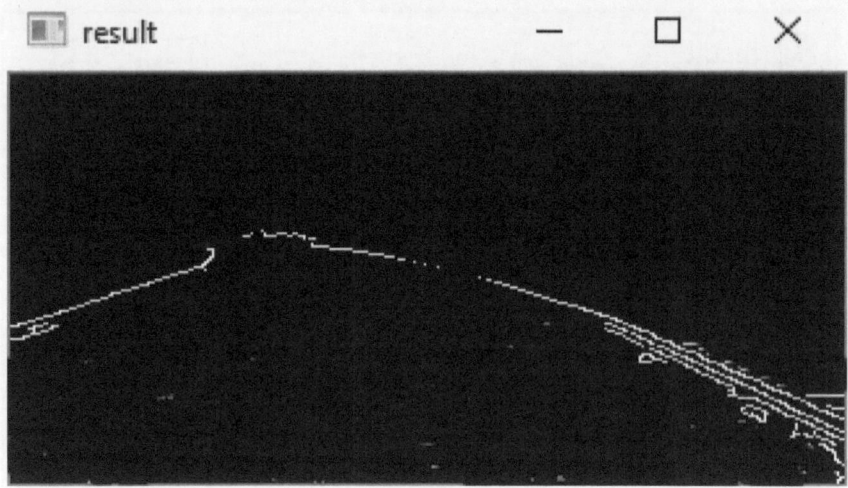

Figure 5.13 Masked image.

This shows any line that has a length of at least 1 in the region of interest and a maximum line gap between two lines in 5; then join the lines to make the lane lines (see Figure 5.14).

Average the lines and remove any excess to generate even-looking lane lines.

$$param = np.polyfit\big((x1,x2),(y1,y2),1\big)$$

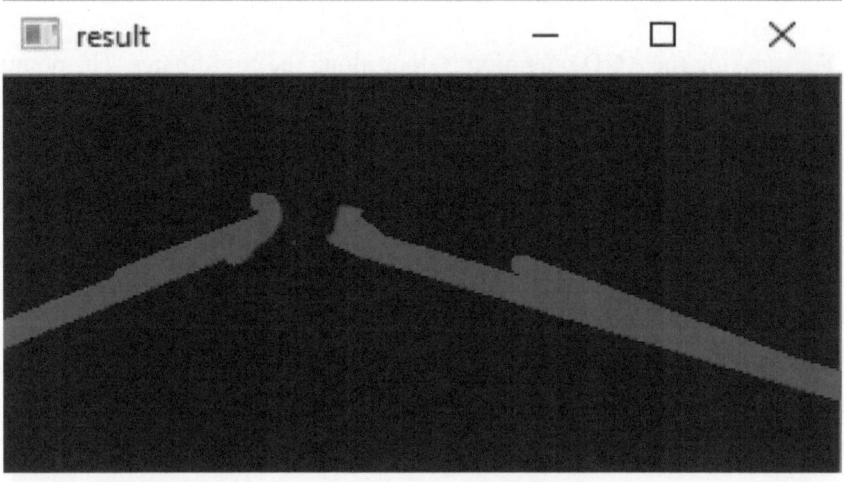

Figure 5.14 Lane lines.

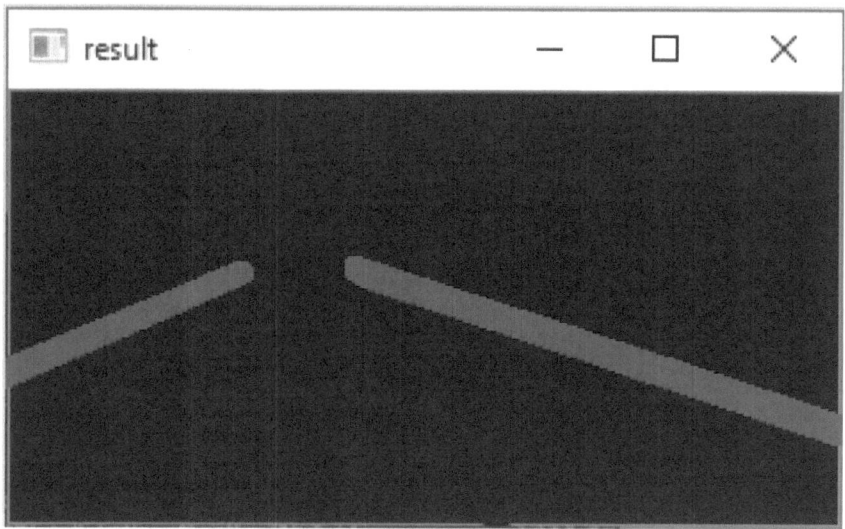

Figure 5.15 Lane image.

Where $(x1, y1)$ and $(x2, y2)$ are the coordinates of the lines

$$left_{fit}.append\big((slope,\ intercept)\big)$$

$$right_{fit}.append\big((slope, intercept)\big)$$

$$left_{fit_{average}} = np.average\big(left_{fit}, axis = 0\big)$$

$$right_{fit_{average}} = np.average\big(right_{fit}, axis = 0\big)$$

The resulting lane image is as follows (see Figure 5.15).

Superimposing the lane lines on the original image leads to the resulting image (see Figure 5.16).

$$combo_{image} = cv2.addWeighted\big(lane_{image}, 0.8, line_{image}, 1, 1\big)$$

Table 5.3 consists of a part of the log table that records the steering angle, throttle, slowing, and velocity. The first column is steering angle; the second column is throttle or acceleration; the third column is slowing or deceleration; and the last column is the velocity of the car.

Figure 5.16 Road image with lane lines

Initially, when the car is not accelerating, the values are 0. Later it picks up speed.

Now we perform a genetic operation on the final image, that is, Figure 5.15, to get the mutated and crossovered lane image and share it on the blockchain framework.

First we convert the final image to grayscale (see Figure 5.17).

After this, we fetch the bits using the aforementioned steps and perform mutation (see Figure 5.18).

After this, we perform crossover using Table 5.3 (see Figure 5.19).

Similarly, the car location is also mutated and crossover is performed using the same approach before transferring it over the network. A few of the results appear in Table 5.4.

The security strength is tested using a plaintext sensitivity test and key sensitivity test on the image and location data. The sensitivity percentage in images is about 55 percent and location data is 60 percent, which leads to the avalanche effect, and hence determines that the security level is optimal.

We have simulated the blockchain framework for the connectivity of smart vehicles in OMNET++ environment. In Figure 5.20, each node represents a smart car connected to the blockchain framework.

Table 5.3 Log table

Steering angle	Throttle or acceleration	Slowing or deceleration	Velocity
0	0	0	9.05E-06
0	0	0	3.40E-06
0	0	0	1.09E-05
0	0	0	1.06E-05
0	0	0	1.15E-05
0	0.363962	0	0.518547
−0.25	0.619169	0	1.088658
−0.45	0.261005	0	1.572785
−0.65	0.019939	0	1.597308
0	0	0	1.574215
0	0.247937	0	1.787757
−0.2	0.509535	0	2.379004
−0.4	0.268125	0	2.623336
−0.6	0.006975	0	2.645121
−0.85	0.252324	0	2.828777
−0.2	0.002654	0	2.804721
−0.45	0.266352	0	2.807959
−0.65	0.530717	0	3.009142
−0.85	0.786221	0	3.002638
−1	1	0	2.978517
0	1	0	2.954589
−0.2	0.734102	0	2.926592
0	0.475661	0	2.895541
−0.2	0.266352	0	3.374302
−0.45	0.530717	0	3.94682
−0.65	0.786221	0	4.632617
−0.85	1	0	5.495206
−1	1	0	6.548716
−1	0.734102	0	7.178916
0	0.475661	0	7.430546
0	0.236282	0	7.595921
−0.2	0	0	7.212229
−0.4	0.262001	0	7.362949
−0.65	0.525987	0	7.844231
−0.85	0.77992	0	8.436392
−1	1	0	9.205722
−1	1	0	10.02749
−1	1	0	11.07054

Figure 5.17 Grayscale of road image with lane lines.

Figure 5.18 Mutated image.

Figure 5.19 Crossover image.

Table 5.4 Mutation and crossover of location coordinates

Location	Mutated location	Crossover location
28.7041° N, 77.1025° E	15E99F85C82988B2490C5B478BC4C05C	CE58670C9B34D5940817725154B85065A3514B8
19.0760° N, 72.8777° E	0E0931EEC777476F37026A57551216AC	5093468 4ED26C1C36B51A3255152F57D03E50DDC
22.5726° N, 88.3639° E	2791347BE359046D9E1B51AB50BF2F69	5B4F2F0020B21FFA2D029098002832EF42BE6BAE
51.5074° N, 0.1278° W	1AB877A8CE6C914B33EAF28709A218A1	3012C78D826AB6CF6970EFB46D1DEF79F1261E2F
23.8103° N, 90.4125° E	EE4377BE7AD0B0AF2ABE71F1E8D7696A	17D64E7E069BF688EEC668F027397829F8B0656A
40.7128° N, 74.0060° W	BAFF956EA9AEC0D5CFE923B3EE9C1660	25832FEFC118FB847264BD96D93960A50F4D014B
25.2744° S, 133.7751° E	960D03A53FCFE9576BE0F48ACC3CC237	5C3ED38BC15DA067F532A6154A98355786E11D54
8.7832° S, 34.5085° E	8E188D18A38C47E2E24DAAFC0CD1E734	D83A32E58167AD742EC1B13A819817BAAE905C19

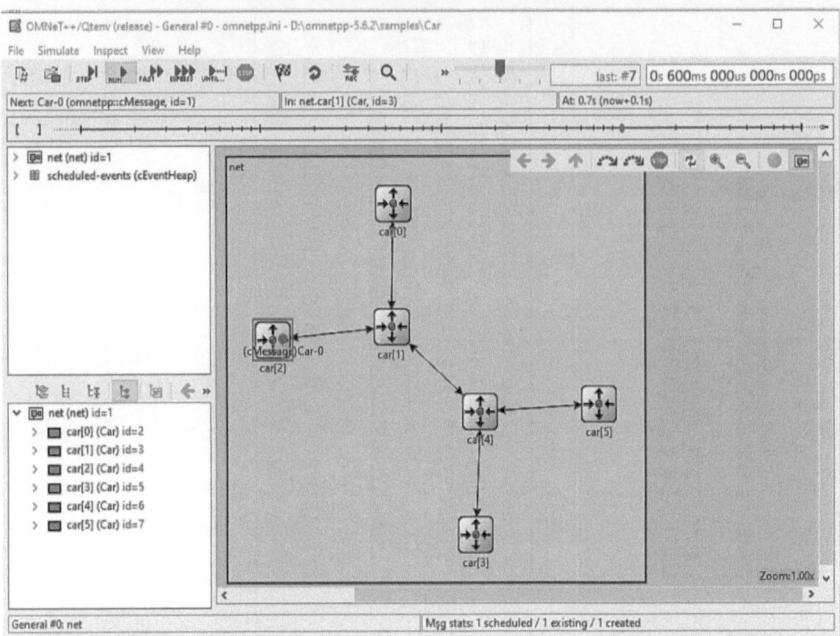

Figure 5.20 Blockchain simulation.

5.4 CONCLUSION

We have proposed a post-quantum blockchain-based security framework that is protecting the privacy of the users and the smart vehicle ecosystem. The algorithm uses a combination of a genetic approach based on natural selection and deep recurrent neural networks. This generates a temporal dynamic network of the blockchain. The security analysis is done based on sensitivity tests implemented in Python. The smart vehicle grid is simulated in OMNET++. The results show optimum sensitivity with an avalanche effect above 55 percent that ensures a secure rental smart vehicle system against quantum computing attack.

REFERENCES

[1] Huang, X., Xu, C., Wang, P., Liu, H., 2018. LNSC: a security model for electric vehicle and charging pile management based on blockchain ecosystem. IEEE Access 6, 13565–13574. https://doi.org/10.1109/ACCESS.2018.2812176.

[2] Goryaev, N., Myachkov, K., Larin, O., 2018. Optimization of "green wave" mode to ensure priority of fixed-route public transport. In Transportation Research Procedia 36, 231–236.

[3] Gorodokin, V., Almetova, Z., Shepelev, V., 2017. Procedure for calculating on-time duration of the main cycle of a set of coordinated traffic lights. In Transportation Research Procedia 20, 231–235.

[4] Chłopek, Z., Biedrzycki, J., Lasocki, J., Wójcik, P., Kieracińska, A., Jakubowski, A., 2014. Examining of the effectiveness of operation of a system to reduce particulate matter emission from motor vehicle brake mechanisms in conditions simulating the real vehicle use. The Archives of Automotive Engineering – Archiwum Motoryzacji 63.1, 35–50.

[5] Skrúcaný, T., Kendra, M., Stopka, O., Milojević, S., Figlus, T., Csiszár, C., 2019 Impact of the Electric Mobility Implementation on the Greenhouse Gases Production in Central European Countries. Sustainability 11.18, Paper no. 4948.

[6] Bukova, B., Brumercikova, E., Cerna, L., Drozdziel, P., 2018. The Position of Industry 4.0 in the Worldwide Logistics Chain. LOGI - Scientific Journal on Transport and Logistics 9.1, 18–23.

[7] Narbayeva, S., Bakibayev, T., Abeshev, K., Makarova, I., Shubenkova, K., Pashkevich, A., 2020. Blockchain Technology on the Way of Autonomous Vehicles Development, Transportation Research Procedia, Elsevier, 168–175.

[8] Gao, F., Zhu, L., Shen, M., Sharif, K., Wan, Z., Ren, K., 2018. A blockchain-based privacy-preserving payment mechanism for vehicle-to-grid networks, IEEE Network.

[9] Jiang, P., Guo, F., Liang, K., Lai, J., Wen, Q., 2017. Searchain: blockchain-based private keyword search in decentralized storage. Future Generation Computer Systems.

[10] Rathee, G., Sharma, A., Iqbal, R., Aloqaily, M., Jaglan, N., Kumar, R., 2019. A Blockchain Framework for Securing Connected and Autonomous Vehicles. Sensors 19.14, Paper no. 3165.

[11] Stopka, O., 2019. Approach technique of specifying a proper autonomous cart type for its service in the logistics center. The Archives of Automotive Engineering – Archiwum Motoryzacji 84.2, 23–31.

[12] Stopka, O., Chovancova, M., 2018. Optimization process of the stock quantity based on a set of criteria when considering the interaction among logistics chain components. In 22nd International Scientific on Conference Transport Means 2018. Trakai, Lithuania, 737–742.

[13] Madhushan, R., Farook, C., 2020. NavAssist-intelligent landmark-based navigation system. In Lecture Notes in Networks and Systems 69, 75–86.

[14] Makarova, I., Pashkevich, A., Mukhametdinov, E., Mavrin, V., 2017. Application of the situational management methods to ensure safety in intelligent transport systems. In VEHITS 2017 - Proceedings of the 3rd International Conference on Vehicle Technology and Intelligent Transport Systems, Porto, Portugal, 339–345.

[15] Tsybunov, E., Shubenkova, K., Buyvol, P., Mukhametdinov, E., 2018. Interactive (Intelligent) Integrated System for the Road Vehicles' Diagnostics. In Lecture Notes of the Institute for Computer Sciences, Social-Informatics and Telecommunications Engineering 222, 195–204.

[16] Khasanov, R., Shepelev, V.D., Almetova, Z., Shubenkova, K., Aristombayeva, M., 2019. Improving computer support system for drivers with multiport memory devices. In VEHITS 2019 - Proceedings of the 5th International Conference on Vehicle Technology and Intelligent Transport Systems. Heraklion, Greece, 663–670.

[17] Ipakchi, A. and Albuyeh, F., 2009. Grid of the Future. IEEE Power and Energy Magazine, 7, 52-62. http://dx.doi.org/10.1109/MPE.2008.931384

[18] Farhangi H., 2010. The Path of the Smart Grid, IEEE Power and Energy Magazine, 8.1, 18–28.

[19] Dong, Z. Y., Luo, F., Lai, J., (2018) Data-centric energy ecosystem in active distribution network. Southern Power System Technology, in press.

[20] Kang, J., Yu, R., Huang, X., and Zhang, Y., 2018, Privacy-preserved pseudonym scheme for fog computing supported internet of vehicles, IEEE Transactions on Intelligent Transportation Systems, 19, 2627–2637.

[21] Samad, T., & Annaswamy, A. M., 2017. Controls for Smart Grids: Architectures and Applications. Proceedings of the IEEE, 105.11, 2244–2261. https://doi.org/10.1109/JPROC.2017.2707326

[22] Danzi, P., Kalør, A., Stefanovi, C., Popovski, P., 2019. Repeat-Authenticate Scheme for Multicasting of Blockchain Information in IoT Systems, arXiv:1904.07069v3.

[23] Singh, M., Kim, S., 2017. Blockchain-Based Intelligent Vehicle Data Sharing Framework, Block-chain based Secure Decentralized Trust Network for Intelligent Vehicles, https://www.researchgate.net/publication/319416415

[24] Singh, D., Singh, M., Singh, I., Lee, H., 2015. Secure and reliable cloud networks for smart transportation services, 17th International Conference on Advanced Communication Technology (ICACT), Seoul, pp. 358–362. doi: 10.1109/ICACT.2015.7224819

[25] Biswas, K., Muthukkumarasamy, V., Singh, K., 2014. An encryption scheme using chaotic map and genetic operations for wireless sensor networks, IEEE Sensors Journal, 2801–2809.

[26] Chowdhury, N., Mackenzie, L., Perkins, C., 2014. Requirement analysis for building practical accident warning systems based on Vehicular Ad-Hoc Networks, IEEE/IFIP 11th Annual Conference on Wireless On-demand Network Systems and Services (WONS), 81–88.

[27] Biswas, K., Muthukkumarasamy, V., 2016. Securing smart cities using blockchain technology, IEEE 18th international conference on high-performance computing and communications, 1392–1393.

[28] Olariu, S., Eltoweissy, M., and Younis, M., 2011. Toward autonomous vehicular clouds, ICST Trans actions on Mobile Communications Computing, 11.7–9, 1–11.

[29] Leiding, B., Memarmoshrefi, P., Hogrefe, D., 2016. Self-managed and blockchain-based vehicular ad-hoc networks. In Proceedings of the 2016 ACM International Joint Conference on Pervasive and Ubiquitous Computing: Adjunct (UbiComp '16). ACM, New York, NY, USA, 137–140.

[30] Banerjee B., Patel J., 2016. A Symmetric Key Block Cipher to Provide Confidentiality in Wireless Sensor Networks, INFOCOMP Journal of Computer Science 15 (1) 12–18.

[31] Banerjee B., Jani A., Shah N., Patel A., 2020, Post Quantum Security Enhancement Scheme for IoT Blockchain Framework, GIS Science Journal 7 (6), 664–672.

[32] Dorri, A., Steger, M., Kanhere, S., Jurdak, R., 2017. Blockchain: A distributed solution to automotive security and privacy, eprint arXiv:1704.00073.

[33] Mihaylov, M., Jurado, S., Avellana, N., Moffaert, K. V., Abril, I. M., & Nowe, A., 2014. NRGcoin: Virtual currency for trading of renewable energy in smart grids. 11th International Conference on the European Energy Market (EEM14). doi:10.1109/eem.2014.6861213

[34] Udacity Self Driving Car Simulator | https://github.com/udacity/self-driving-car-sim

[35] Kaur, J., Saini, J., 2014. On classifying sentiments and mining opinions. International Association of Scientific Innovation and Research (IASIR), IJETCAS 14-580; 2014, IJETCAS.

[36] Kaur, J., Dharni, K., 2019. Predicting Daily Returns of Global Stocks Indices: Neural Networks vs Support Vector Machines. Journal of Economics, Management and Trade. https://doi.org/10.9734/jemt/2019/v24i630179

[37] Park, J., Park, J., 2017. Blockchain Security in Cloud Computing: Use Cases, Challenges, and Solutions, Symmetry 2017, 9, 164; doi:10.3390/sym9080164

An overview of the autonomous vehicle system, security, risks, and a way forward

Md Aminul Islam and Sarah Alqhtani

Oxford Brookes University, Headington, United Kingdom

CONTENTS

DOI: 10.1201/9781315110905-6

6.1 INTRODUCTION

Throughout the years, autonomous vehicles (AVs) have been in constant development. AV technology is not a new technology, but since the 1990s, initiatives for innovation have existed in the private and public sectors. These innovations include many systems for advanced driver assistance and some functional AVs [1]. This technology creates a potential impact on travel behaviors, congestion, and safety. Other social impacts this technology provides include fuel efficiencies, parking benefits, reduced travel time, and crash savings, which is estimated to be $2,000 per year per AV [2]. There are some challenges that may occur with AV, such as system performance in complex, heavily-trafficked environments and safety with uncertain interactions with other participants in traffic. Questions regarding reliability and safety need to be addressed, which can occur in the process of end-to-end learning and interactive planning. However, one of the most important purposes of the AV is to keep passengers and traffic safe. Most accidents happen due to human error, and on a yearly base, more than 3,000 lives are lost daily [3].

In this chapter, the AV system is discussed in detail. There are many sensors, technologies, and algorithms that can be used to create different models of AV. However, within any technology, some issues and risks will arise. Issues vary from social, professional, environmental, legal, and ethical, and will be discussed in conjunction with risk management assessment.

6.2 AUTONOMOUS VEHICLES

AVs are usually not organized systematically. The vehicle system can differ in so many ways. Sensors are usually very different from one design to another. Some designs completely depend on cameras, while other designs can also include a combination of technologies such as laser scanners, milliwave radars, Global Positioning System (GPS) receivers, and cameras. In AVs, software issues should be addressed due to the large-scale platform. It is more efficient to build up the platform from scratch. The design process and application of algorithms such as path planning, vehicle control, and scene recognition need a significant amount of integrated collaboration of knowledge and skills, primarily engineering skills. Furthermore, the use of algorithms in AVs demands huge databases to train the recognition process, and, most importantly, for localization maps of public roads [4].

6.2.1 Sensors

In every AV the main component that makes it autonomous is sensors. The fusion of data gathered from the sensors that should be accurately interpreted, along with the vehicle control system is the most important point in AV. For better visualization, the process of the AV system can be divided into four main categories (see Figure 6.1). The first category is the sensors, which can be many different kinds that are installed in the vehicle. These hardware components are installed in the system to sense the surroundings of the vehicle. After that the sensors gather the data to translate and process them into meaningful information, which is the perception step. Furthermore, the output of the perception process is used to determine behavioral planning for the long-range and the short-range path plan. Then there is the control system that takes the path plan that is generated from the planning subsystem and sends to the vehicle control commands accordingly.

Sensors are a very important part of the AV, and different types of sensors are used in AV designs. The main types of sensors that are most commonly used in this system are radar, cameras, sonar, wheel odometry, Lidar (light detection and ranging), inertial measurement unit (IMU), and GPS. These sensors are used to collect data, process them in the impeded computing system, and use them to control the speed, brakes, and steering of the vehicle.

6.2.1.1 Camera

Camera sensors are typically the preferred choice of sensors used in manufacturing an AV, and they are the first type of sensor to be used in this system.

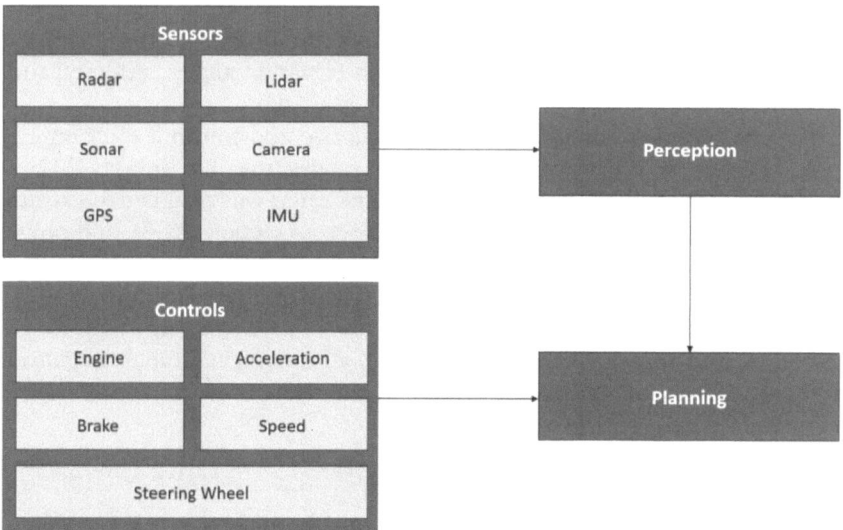

Figure 6.1 Autonomous vehicle system diagram.

They are the best way to visualize the surroundings of the vehicle. There are many reasons why this type of sensor is commonly used, mostly because it is affordable, available, and can interpret texture accurately. However, the camera needs computational power to process the data. In recent high-definition cameras, the pixel per frame rate can reach up to a million, and this also includes 30 to 60 frames per second. This means that for a camera to process in real time, it processes a multimegabit of data [5].

Unlike the other sensors, which are active, the camera, or the optical, sensor is a passive sensor. It is called a *passive* sensor because it gathers data in a way that is non-intrusive. Due to the affordability of a camera, vehicles can have it both in the rear end and the front end of the car. This provides a 360-degree view that can make tracking cars coming from around a curve or switching lanes more efficient. Moreover, there are many applications that use visual information, such as traffic-sign recognition, lane detection, and object identification, which can vary among pedestrians, obstacles, and many other objects that can be found on the road. These applications do not require modification within the road infrastructure [6].

6.2.1.2 Radar

Radio Detection and Ranging (Radar) is a sensor that is embedded in the vehicle for different applications like collision warning and avoidance, adaptive cruise control, and blind spot warning [5]. These applications are achieved by using radar in object detection, determining their position, and detecting the speed relative to the vehicle. The progressed development in millimeter-wave semiconductor technology and signal processing techniques are a main factor in a radar system. The techniques in signal processing make the resolution and estimation process better in every measurement dimension, including azimuth-elevation angles, velocity, and range of the surrounding objects around the vehicle.

The radar sensor has the capability to receive and transmit electromagnetic (EM) waves. The wave frequency band ranges from 3 megahertz (MHz) to 300 gigahertz (GHz). As with other sensors, radar was designed to extract information such as velocity, location, radar cross section (RCS), and range using the reflected wave from the object targeted. The AV radar system operates at the frequency band 24 GHz and 77 GHz in the EM spectrum, which is called *mm-wave frequency*. These frequencies are used to achieve high-range resolution and velocity. Three main tasks of a fundamental radar operation are range estimation, direction estimation, and relative velocity.

6.2.1.2.1 Range estimation

To get the range estimation of a target, the range R can be calculated based on the time delay of the round-trip of the EM waves. It is extracted from the target: $R = (c\tau/2)$, where the delay time of the round-trip is determined in seconds by the value τ, and the value c is the speed of light in m/s (c≈3×108 m/s). This τ value estimation will determine the range measurement. The radar transmits a type of EM waves that is necessary for the time-delay estimation of the round-trip. One example is the pulse modulated continuous waves (CWs), which contains short and periodic power pulses and silent periods. The silent period enables the receiving process of the radar to receive the reflected signals, and it also serves as a timing mark for the radar's performance of the range estimation. However, CW signals that are unmodulated, such as cos(2πfct), lack those timing marks; therefore, they are unused for range estimation. Furthermore, the reflected signal should arrive from the target before the start of the next signal. Hence, the radar's maximum detectable range relies on the pulse repetition interval transcutaneous pulsed radiofrequency (TPRF). Due to the loss and imperfection of the reflection path from the target, the signal that is transmitted and received from the radar undergoes attenuation. The radar electronics have internal noise that can affect the received target signal, alongside interferences that are caused by the reflected signal of other objects and human-made sources such as jamming. The ambient noise in the form of additive white Gaussian random process is the only thing that is considered in a standard round-trip time delay estimation. The demodulation is assumed to already remove the carrier for the target signal x(t) at baseband to be modeled as:

$$x(t) = \alpha s(t - T) + \omega(t),$$

The value of α is a complex scalar that has a magnitude that represents attenuation that is caused by path loss, RCS of the target, and antenna gain. The $\omega(t)$ represents the additive white Gaussian noise, which has variance σ^2 and zero mean. The goal here is to estimate the τ using the knowledge of the radar waveform that is transmitted s(t). With the assumption of finite energy Es and having a unit amplitude in the signal s(t), a radar receiver that is ideal can be using a matched filter with the impulse response h(t)=s*(−t). That maximizes the ratio of signal to noise (SNR=(α^2Es/σ^2)=(α^2Tp/σ^2)) in the output. The correlation between the transmitted and the received signal pulses in this case is found by the matched receiver, which is filter based:

$$y(T) = \int x(t)s^*(t - T)\, dt.$$

The time delay maximum likelihood (ML) is estimated to be the time that the magnitude of the output's matched filter, which peaks at:

$$\hat{\tau} = \text{argmax}\tau\, |y(\tau)|.$$

The existence of the noise can perturb the peak's location, which will give an estimation error. The radar also needs to determine if the received signal has an echo signal from the target. Many strategies are being studied and developed when it comes to radar technology, but most widely recognized decision strategy can be formed by statistical hypothesis testing, which leads to a simplified threshold testing in the matched filter output. Another main performance measure is range resolution, which distinguishes closely spaced targets. If a nonoverlapping return signal in the time domain is produced, two targets can be separated in the range domain only. Therefore, the pulse width Tp is proportional to the range resolution, which means that higher pulses give higher resolution. Less energy can be found in shorter pulses, which means it has a poor receiver detection performance and signal to noise ratio (SNR).

6.2.1.2.2 Direction estimation

With a wideband pulse like the frequency modulation continuous waves (FMCW), a discrimination of the target's velocity and distance is provided. The direction's discrimination is enabled by antenna array. With traffic, for example, there are usually several targets in the area surrounding the radar, which collect reflections that are direct and multipath. An angular target's location is estimated to deliver a comprehensive representation of the traffic scenery. Thus, in automotive radars, the description of the target's location is in the terms of a spherical coordination system (R, θ, ϕ), where the (θ, ϕ) coordination denotes respectively the azimuthal and elevation angles. If a single antenna radar is used in this case, in problems of range-velocity estimation, it might not be sufficient because of the time delay measure $\tau = (2(R \pm vt)/c)$, which lacks the angular locations information of the target.

To make the direction estimation more feasible, the data of the reflected signals along multiple distinct dimensions should be collected by the radar. when a target is being located using EM waves in 2-D, the data of the reflected wave from the target should be collected in two distinct dimensions. Those two dimensions can be a combination of space, time, and frequency. For example, in a wideband waveform such as FMCW and in antenna arrays, two unique dimensions will be formed. However, an mm-wave band, which is a smaller wavelength, correlates with smaller aperture sizes. Therefore, numerous amounts of antenna elements can be packed densely into an antenna array. Hence, a stronger and sharper beam, which is the effective radiation beam, in turn, increases the angular measurements' resolution.

6.2.1.2.3 Velocity estimation

Based on the Doppler effect, the velocity of a target estimation can be found. Assuming that a vehicle is moving forward in a differential velocity (v). considering the relative motion between two vehicles, there is a delay in the reflected waves by time $\tau = (2(R \pm vt)/c)$. The received wave will experience a frequency shift known as the *Doppler shift* $fd = (\pm 2v/\lambda)$, which is caused by the time dependent delay term. Depending on the direction of the target, whether it is moving away or approaching, the sign is determined to be negative or positive. This sign along, with the wavelength λ, are inversely proportional with the Doppler shift. Although the shift in frequency is possible to detect using CW radar, it is not able to measure the range of the target.

In the case of FMCW radar, it transmits a periodic wideband FM pulses that has an angular frequency that increases linearly during the pulse. For the FM modulation constant K and the carrier frequency fc, the single FMCW pulse is shown as:

$$s(t) = ej2\pi(Jc + 0.5Kt)t0 \leq t \leq T.$$

The reflected signal from the target is combined with the transmitted signal to generate a beat signal with low frequency, which gives the range of the target. The process is repeated for P consecutive pulses. Waveforms that are two-dimensional represent a reflective pulse that is arranged successively in two-time indices. The pulse number corresponds with the slow time index p.; however, the fast time index n is assuming that for every pulse, the continuous corresponding beat signal is sampled with the frequency fs to get samples N during the duration time T. If the single target is assumed and the duration of the reflected signal is neglected, the radar receiver FMCW output function of the two-time indices is shown as:

$$d(n,p) \approx \exp\{j2\pi[(2KRc + fd)nfs + fdpT0 + 2fcRc]\} + \omega(n,p).$$

A discrete Fourier transform through fast time n is applicable to get the beat frequency fb=(2KR/c) coupled with the Doppler frequency fd. Another name for this operation is *range gating* or *range transform*. This operation is able to estimate the Doppler shift that corresponds with a unique gate range, which is possible by applying a second Fourier transform along the slow time. Using the two-dimensional Fast Fourier Transform (FFT) will enable finding the range-Doppler map [7].

6.2.1.3 Lidar

Lidar sensors operate on infrared (IR) laser beams to estimate the distance between the sensor and an object. Lidar sensors that are currently used operate on light in the 900 nm wavelength range. However, longer wavelengths are used in some Lidar sensors, which provide better performance in harsh conditions such as rain, snow, and fog. The Lidar uses laser beams in pulses across its field view, which are reflected by the objects within that view. It also has the capability to detect objects at a range that varies from a few meters to more than 200 m. Yet, unlike the radar sensor, it may face difficulties in detecting objects that are at a very close distance. This sensor has a better spatial resolution than the radar sensor. That is because the Lidar has a larger number of vertical direction scan layers, denser Lidar points per layer, and a more focused laser beam. It is able to collimate a laser beam with its short 905 to 1,550 nm wavelength; therefore, it is able to make IR spatial resolution on the order of 0.1 degrees. This makes extremely high-resolution 3D characterization of an object within the scene of the sensor possible, without any significant back-end processing. When it comes to detection in a low-light condition, at night, for example, Lidar sensors can perform very well. Having said that, the Lidar sensor is lacking when it comes to directly measuring the velocity of an object and it must depend on various positions between two or more scans. It is also affected by the dirt on the sensor and weather conditions in general [5, 8].

There are many types of Lidar systems that can be applied to an AV. For the narrow-pulsed Time of Flight (ToF) method, there are two types in Lidar beam steering systems. The first type is the mechanical Lidar, which uses a rotating assembly and high-grade optics to create a wide Field of View (FOV), usually 360 degrees. This aspect provides a high SNR within a wide FOV. Nevertheless, it generated a bulky implementation result. The second type is the solid-state Lidar. This Lidar has a reduced FOV and no spinning mechanical components. Due to these features, this sensor is cheaper than other types. To make a sensor that has an FOV that can compete with the mechanical Lidar, multiple channels distributed in the front, the rear, and the sides should be applied to the vehicle, while fusing their data [8].

For the mechanical Lidar, also known as *spinning Lidar*, the system has an IR-coherent light that is emitted from the laser. After that, the light is circularized and collimated to a round bean using optics. Every beam is put with a matching receiver, usually an avalanche photodiode (APD). Several pairs of the emitter detectors are mounted on a column that uses a motor to spin it, usually in frequency value between 10 and 20 Hz. The cycle's duty is low to guarantee eye safety. The number of pairs of emitter detectors stacked up vertically determines the vertical FOV, while the speed of the motor's rotation and cycle duty determines the resolution of Horizontal FOV. This system is known to have the cleanest signals, a ratio of noise to date, and provides a 360-degree HFOV. However, the cost and size are a challenge in this system. There are also some doubts about the need for self-calibration because of motor bearing wear [9].

Several methods for solid-state Lidar implementations are discussed next.

6.2.1.3.1 Microelectromechanical system (MEMs) Lidar

The MEMs Lidar system implements very small mirrors with a tilted angle that vary when a stimulus like voltage is input. This system substitutes an equivalent of the mechanical scanning hardware, which is electromechanical. Typically, in this system, the light-collection receiver aperture, which determines the SNR received, is small, moving a few millimeters. Cascading several mirrors is required for the laser beam to move in several dimensions. The alignment process is not significant; however, when installed, it has a higher risk of vibrations and shocks, which are commonly experienced in a moving vehicle.

6.2.1.3.2 Optical phase array (OPA)

The OPA system is almost equivalent to the phased array radar. It contains an optical phase modulator that controls the speed of light going through the lens. This speed of light control process makes the optical wave-front shape control possible. A phenomenon occurs that would effectively steer the point of the laser beam to a different direction. It happens when the top beam does not experience any delays, but the other beams have delays in an increasing amounts. Other methods are also able to steer the backscattered light in the direction of the sensor, which eliminates the mechanical parts.

6.2.1.3.3 Flash Lidar

The operation in the flash Lidar is similar to the one in a standard digital camera, which applies an optical flash. A single pulse from a large-area laser illuminates the surroundings in front of it and a photodetectors focal-plane array placed close to the laser captures the back-scattered light's proximity. The image location, distance, and reflected intensity is captured by a detector. Given with a single image, the entire scene is captured, unlike with the mechanical laser scanning method, it has a faster data-capturing rate because the entire image is captured in a single flash. In this method, the system is more immune to the vibration effect that can tamper with the image. However, the method has a retroreflector that back-scatters a bit of the light and reflects most of it, which, in a real-life environment, blinds the sensor entirely and makes it useless. Another downside is its high-peak laser power, which is necessary to detect from long distances and to illuminate the whole scene.

6.2.1.3.4 Frequency-modulated continuous wave (FMCW) Lidar

Unlike the other methods mentioned, this method does not use the ToF principle, which uses narrow light pulses. The FMCW Lidar adapts to the coherent method that produces a small chirp of frequency modulated laser light. The system is able to measure velocity and distance by measuring the frequency and phase of the returned chirps. The optics and the computational load are much simpler with the FMCW Lidar, yet generating the chirps adds complexity [8].

6.2.2 Detection and algorithms

A deeper understanding of the environment is needed for the AV to oper-ate and react correctly. The roads in that environment are very hectic and unpredictable. The system should recognize objects and their locations. The objects are usually vehicles, pedestrians, and bicycles. There are also stationary objects on the road that the AV should recognize, such as traffic lights and signs. The AV uses algorithms to classify its components. Some of those components are path planning, scene recognition, and vehicle con-trol. Those classes contain a set of algorithms such as object detection, object tracking, and scene recognition that requires localization. The path planning, for example, usually falls into motion and mission planning, whereas the vehicle control falls into path following.

6.2.2.1 Localization and mapping

The AV needs a higher accuracy in navigating than the available GPS guid-ance systems, especially when navigating in a dynamic urban environment. The precision in localization determines the reliability of autonomous driving. The algorithm used to enable this localization problem is the Normal Distributions Transform (NDT) algorithm, and in particular the 3D module of it. The 3D NDT version applies a scan matching over 3D map data and 3D point cloud data. This results in a localization that can perform in an order of centimeter, providing a high-precision 3D map and high-quality 3D Lidar leverage. Not only can the NDT algorithm be used in 3D form, but it also has a computational cost that is not affected by the map size.

However, a simple way of applying localization is by obtaining data that helps determine the vehicle's location through the observation of known and fixed beacons (points of reference) in an active environment. It is very straightforward in principle to get the position by giving a number of bea-con observations. AV uses a wide range of systems that are beacon based for the localization problem. In specific, the factory AV use of the laser system and reflective beacons is very common. The beacons are based on ultrasonic transducers, radar reflectors, and microwave tags [4, 10].

Another algorithm that is used to help with the localization problem is the Kalman Filter (KF) algorithm. This algorithm is proven to be able to deal with complex localization problems in a very simple way, yet it has many limitations. In the KF algorithm, an estimate of the point of inter-est is made by a recursive linear estimator, which utilizes a model of the observation of the process and a model of the process under consideration. It is common to start the localization problem with a discrete time process model, which describes the platform's motion as:

$$x(k) = f\big(x(k-1),\, u(k)\big) + w(k)$$

The state or location of the vehicle at a time k is described as x(k), and the average input or drive signals of the vehicle at a time k is described as u(k). The f is the kinematic function, which describes the vehicle's transition in the time between k-1 and k. While the w(k) presents the value of a random vector that describes the motion uncertainty of the vehicle in time. Observation of the vehicle's location at a certain time is made, typically by getting the location of the beacons measurements that are applied in known places around the environment, according to the observation equation:

$$z(k) = h(x(k)) + v(k)$$

The z(k) presents the observation that was made at time k, and the h presents a function relating the current location of the vehicle to the observation. The v(k), however, is described as a random vector of the error that occurs in the observation process. The KF computes an estimated vehicle state recursively through a combination, which is linear, of the prediction based on the process model:

$$x(k \mid k-1) = f(x(k-1 \mid k-1), u(k))$$

However, the correlation that is based on the difference of prediction and true observation is shown as:

$$x(k \mid k) = x(k \mid k-1) + W(k) \, [z(k) - h(x(k \mid k-1))]$$

The W(k) is the gain obtained by the observation and the relative confidence in the prediction of the vehicle.

The KF algorithm offers many features, including a continuous confidence in the estimate of the location through the matrix of state of covariance, the ability to estimate a none directly observed state, the ability to describe a variety of sensors in a single coherent form, and the incorporation, incrementally, and recursively, of information into the estimate of a vehicle's location (see Figure 6.2). Furthermore, the KF algorithm also has some limitations to consider. The first is the need for accurate models of both observations and vehicles. Also, linearization assumptions with a linear predictor corrector form of the updated equation are necessary. Also, the terms of the true noise should be bound by the first two moments of the assumed noise model and it should be well behaved. The last limitation is the form of the model applied, which is an analytic form, which makes capturing many physical environments and sensor properties difficult [11].

Figure 6.2 Visualization of the scanning process. The Lidar scanner acquires the IR ground reflectivity and range data. Thus, the resulting map is 3D IR images of the ground reflectivity [12].

6.2.2.2 Object detection

The object detection algorithm is used in AV to avoid accidents and abide by the traffic rules. Specifically, it is used to recognize vehicles, pedestrians, bicycles, traffic lights, and traffic signs. Many features are desired in an object-detection system, such as accuracy, real-time capabilities, and the capability to learn without the need for massive hand-labeled training sets, as well as a system that is inherently multi-descriptor, inherently multi-class, requires less manual feature engineering, and can add new object classes and descriptors without having to relearn from scratch.

To subtract deeper information, the images are segmented. This results in the removal of the local ground plane and the components, connected in clusters, of the remaining points, which are applied on a 2-D grid for efficiency. The segment is input into a standard tracker that includes velocity and position in its state variable. This track classification is important for high performance. Two boosting classifiers are applied to achieve track classification. One is the motion descriptors of the whole track, and the other is the object's shape in each frame of the track. A Bayes filter is used to combine these predictions, and it is shown as the frame descriptor in Figure 6.3 for three objects: a pedestrian, a car, and a bicyclist [11, 13].

When it comes to object recognition, there are two methods: the supervised method and the semi-supervised method.

6.2.2.3 Supervised method

The system is real time and accurate with good tracking and segmentation, and is inherently multi-descriptor and multi-class. The laser-based

Figure 6.3 Recognized virtual camera intensity images for three objects tracked. The segmentation depth of the object provides invariance to the background clutter. The depth classification data enables objects to be added in a canonical orientation, therefore providing some perspective invariant measures [10].

objection detection algorithm on junior has three components, which are segmentation, tracking, and track classification. In the first step, the object gets segmented from the surrounding environment using depth information. After that, it gets tracked with a KF. Therefore, the tracking and segmentation methods are model free, which means that no object class models are applied during the stages. The tracked object's classification is achieved when applying a boosting algorithm across a few high dimensional descriptors spaces, which encode shapes, motion properties, and sizes [14].

The largest error source in the system can be found in the segmentation and tracking process. Objects like bicycles and cars continuously avoid being segmented together with the surrounding environment, yet when it comes to pedestrians and other objects of interest, this is not the case. When an object gets closer to another object, the system generates a false negative.

In the meantime, to maintain a real-time capability, the system needs to remain feed-forward, because more mathematically complex methods that simultaneously take segmentation, tracking, and classification into consideration do not have real-time capabilities. For example, when it only classifies pre-segmented objects, the system takes more time processing each candidate than the sliding window system prevalent in the computer's vision.

6.2.2.4 Semi-supervised method

Segmentation and tracking that is model free creates highly effective learning object model methods that disregard the need for huge numbers of hand-labeled data. This is called *tracking based on a semi-supervised method.* This method actively learns a classifier and collects new and useful training instances by applying tracking information. In bicycles, for example, the method can learn to identify half-occluded bicyclists from tracks that are unlabeled, which includes views that are half-occluded and occluded. The method had been tested to show a relatively high accuracy of track classifications with three hand-labeled training tracks for every object class only [13].

6.2.3 Cloud robotics

The term *cloud robotics* was introduced by James Kuffiner from Google in 2010. This term illustrates a new method that allows robotics to use the internet as a source for large parallel computation and exchanging massive data resources in real time [15]. It enables on-demand access to an almost unlimited computational resource, which can come in handy when dealing with bursty computational workloads that need large amounts of computation periodically. The idea of a remote computer in robotics in general is not new, but it allowed for so many different possibilities for applications in mobile robot systems like the AV. The AV system can access large-scale map data and images through the cloud, which eliminates the need for local data storage. It also gives the AV systems the capability to communicate with each other. However, some challenges come up when using remote cloud resources, particularly when using commodity cloud facilities like Microsoft Azure and Amazon Web Services. Those services may bring some variables that are beyond the robotic system's control. There is also the issue of communicating with remote clouds, which can cause unanticipated delays in the network. The time of cloud computation can also depend on available computer resources and the number of jobs running in that system at the time. In that case, even though the cloud can achieve real- time performance in a normal case, cases when it may get overloaded, latencies may occur and affect the onboard processing needed for critical tasks. However, target recognition can be moved to the cloud, while maintaining the stability control, short-term navigation, and low-level detection local. Using this hybrid method can lower the costs of detecting many objects in near real time, while limiting the negative consequences when the target cannot be met in real time [16, 17].

This method is applied in vehicles like the Google autonomous driving project. The system is enabled to access images and maps that are collected and then updated into the satellite, crowdsourcing, and street view from the network, which generates accurate localization. Another example for this application is the Kiva System, which uses a new approach to warehouse logistics and automation by applying a huge number of mobile platforms to move pallets by updating tracking data and coordinating planforms using a local network [15].

6.2.4 Internet of Things (IoT)

An integral part of IoT is large-scale application of interconnected actuators and sensors. Many applications are made possible with IoT technology, including smart transportation systems. Systems that apply communication, control technologies, and computing together are called *cyber-physical systems* (CPS). They require real time processing, because the sensor network is integrated with the control of the physical system. The technology provides a system where vehicles can communicate. It can be used in intersection management applications. However, this application requires secure, low-latency, reliable, and short-range communication [18].

After the invention of IoT, an emerging concept came called the Internet of Vehicles (IoV). It was the result of the advancements made in artificial intelligence (AI), wireless networks, and sensor technology. This technology already established architectural proposals, including Cisco. The Cisco architecture includes four layers. The first is the end point, which contains the vehicle's software and hardware. The second layer is the infrastructure, which defines all technology that enables connections. The third layer is the operation, which monitors the flow-based management and the policy enforcements. The last layer is the virtual layer, which contains all different types of cloud subscriptions [19].

6.2.5 Blockchain

In a scenario where AVs are part of an accident, how are accidents recorded to determine the cause and liability? When it comes to recordings, it is important the data is trusted, verified, and not tampered with. To achieve these properties, blockchain technology can be used for event-recording schemes, which can be useful for the forensics system. The blockchain contains a series of blocks, and each block has a set of transactions that are timestamped and a hash of the previous block [20]. Originally, blockchain technology is made for *bitcoin*, which is a digital cryptocurrency.

Recording the accident as a timestamp transaction means it is saved in real time into a new block. AV has the capability to save the vehicle from accidents in real time; however, it is not always possible due to the complexities of solving the hash puzzle [21]. A hash function takes a set of inputs of different arbitrary sizes, then puts them into a data structure, such as tables, that has elements with fixed sizes [22]. One of the recording mechanisms to save events into a block is the Proof-of-Event with Dynamic Federation Consensus. In an accident, vehicles that are directly involved broadcast an event generation, which can be via IEEE 802.11p Dedicated Short Range Communication (DSRC), for example, and only the vehicles that have DSRC communication range can respond and receive it [21]. DSRC (is an 802.11p based wireless communication technology that provides direct communication between the surrounding infrastructure and vehicles with high speeds and high security without interrupting the surrounding cellular infrastructure [23]. After that, both involved vehicles receive a command to generate and broadcast the event to the vehicular network that is defined by the existing cellular network infrastructure. A random federation group in

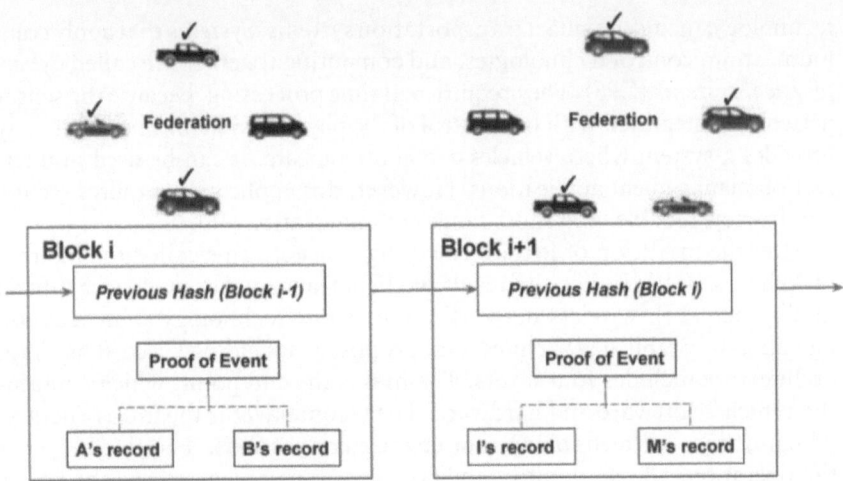

Figure 6.4 The blocks contain hash values and data from the previous block, and it is all confirmed by verified federation vehicles [21].

the vehicular network is created to save and clarify the accident's data into a new block (Figure 6.4). This is done using a multi-signature scheme. The last steps are saving and sending this block to the Department of Motor Vehicles (DMV) to be saved in the permanent record. This DMV mechanism uses the accident's data collected from different sources and the generated hash digest to protect data trustworthiness and integrity. The recordings also provide traceable evidence [21].

6.3 EXISTING MODELS OF AUTONOMOUS VEHICLES

The myth or science fiction of self-driving cars is now a reality with a fast-growing market where riders will enjoy hands-free feeling in a safer way. The market will increase from a $54.23 billion industry to a $556.67 industry by 2026 and 33 million cars will be on road by 2040. [30, 31]

Some car companies are already manufacturing AVs, and some are going to enter the market, such as Tesla, Nuro, Zoox, Baidu, Nissan, General Motors (GM), Aptiv, and so on (see Table 6.1)[32]

Table 6.1 Some AV companies

Name	Cruise	Waymo	Voyage	Nodar	Seeva	ABB	Reality AI
Founded year	2013	2009	2017	2018	2016	1988	2015
Home base	San Francisco	San Francisco	San Francesco	Boston	Seattle	Colorado	New York

6.4 TRAFFIC FLOW PREDICTION IN AUTONOMOUS VEHICLES

With increases in the standard of living, a greater need for the transportation has occurred. There are a higher number of transportation users due to the growing population. This has resulted in more traffic on the roads and more congestion, which is causing a huge problem. Intelligent transportation system (ITS) technology can be applied to solve the traffic problem effectively. ITS merges different technologies, such as artificial intelligence, image analysis, global positioning, electronic information, and many more. Furthermore, ITS technology is considered to be an effective way to solve transportation problems. It also reduces traffic pollution and accidents, and solves traffic congestion in a non-traditional way Two components of the ITS that are of great importance are traffic control and guidance [24]. There are also two main technologies that guarantee a traffic flow guidance system will operate well. The first is an accurate traffic prediction of the data stream, and the second is traffic prediction techniques. A traffic flow is time varying, random, and nonlinear, which makes it hard to predict long-term patterns.

In urban areas, the traffic flow system has characteristics that are chaotic. Taking this idea into consideration, the approximate restoration of the original system prediction is established by a nonlinear mapping construction, which establishes a predictive model. There is various research that has been conducted to establish this model, including methods such as the K-nearest neighbor (KNN), Bayesian network, radial basis function (RBF), neuro-fuzzy system, type-2 fuzzy logic approach, binary neural network, Kalman state space filtering, autoregressive integrated moving average, and many other models. The neural network in these prediction models is the main focus for a lot of experts. This is because neural networks have characteristics, such as distributed storage and massive parallel structure, that are unique. They also have good self-adaptability, self-organization, incorrectness tolerance, the capability to classify patterns, and a powerful function approximation. However, the performance of the network can be affected by improper preferences. The selection of width value and center vectors of the neural network's hidden nodes impacts the generalization and learning abilities greatly [25].

Another algorithm that can be used is the Artificial Bee Colony (ABC). The algorithm can be used for the parameter's combination optimization of the neural network, due to its smaller control parameter and simple structure, easy implementation, and high coverage speed. However, the algorithm is also prone to premature coverage and slow coverage speed.

The ABC algorithm is inspired by honeybee behavior, and the algorithm is assumed to be a colony that has groups of onlookers, employed bees, and scouts. Employed artificial bees make one half of the colony, and the onlookers make the other half. There is one employed bee for every food source, which means that the number of food sources should be equal to the number of employed bees. The scout, on the other hand, is a former employed bee that left the food source. In general, the employed bee determines the food source from the neighborhood memory, then it forwards this information to the onlookers, which can also have one food source. The onlookers have food sources from their chosen neighborhood. After the employed bee leaves the food source and becomes a scout, it starts looking for a new food source. The algorithm can be summed up in four main steps:

1. Determine the nectar amounts of the employed bees and move them to their food source.
2. Determine the nectar amounts for the onlookers and move them to their food source.
3. Move the scouts so they can search for new food sources.
4. Store the best food sources into the memory.

The food source represents possible problem solutions to be optimized. The quality of the solution is represented by the nectar amount. A method called *roulette wheel selection* is applied to put onlookers on their foods. Conversely, the scouts do not have any guidance when looking for food. They just need to find any kind of food source. This behavior causes reduced food sources and low search costs. Scouts sometimes stumble on unknown and rich food sources. The classification of scout bees is controlled by limit control parameters. The limit parameter is equal to the number of trials for releasing a food source, which is a significant control parameter of the ABC algorithm [26, 27].

6.5 CYBERSECURITY RISKS (ISO 27005)

The AV must work in an integrated network system where it will be connected to internet through GPS, cloud, General Packet Radio Service (GPRS), or satellite. However, the networking system must contain a traditional infrastructure, such as a data center, servers, databases, and so on. It is also provisioned with Internet Protocol (IP) addressing and subnet system. The whole system will contain several types of data, such as travel history, maps, rider's data, owner's profile, and so on. So it must comply information security where information risks will arise. To resolve this, we will cover Information Risk Management (ISRM). The subsequent topic focuses on network design and ISO 27005.

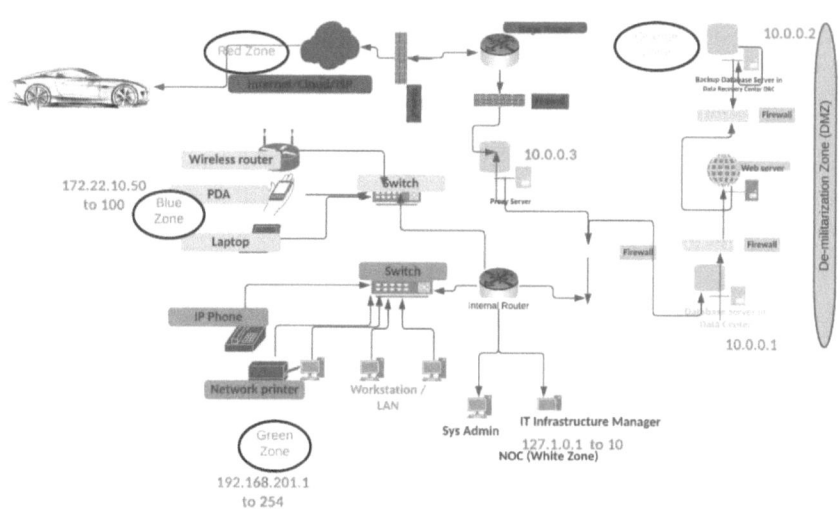

Figure 6.5 Model network design with IP address and zone (by Lucid Chart).

6.5.1 Network design

Figure 6.5 illustrates that the entire autonomous vehicle system must work according to a secure network architecture. The whole network should be in an enterprise where no security compromises are ever permitted. Both of the Network Operation Center (NOC) and Security Operation Center (SOC) should be coordinated from a central hub. All possible aspects of zoning, the proxy server, IP addressing, the router, the switch, and the firewall security system where all modern encryption–Secure Sockets Layer (SSL), Transport Layer Security (TLS), Internet Message Access Protocol (IMAP), Simple Network Management Protocol (SNMP), and Post Office Protocol (POP)–secure protocols would be incorporated as the whole system deals with the lives of humans while they are on the road. Therefore, security is a prime concern.

6.5.2 Risk management (ISO 27005)

Almost all companies who consider their growth in terms of positive branding, sustainability, and customer orientation, place security first. That's why organizations are coming forward to abide by the international standard guidelines. The International Organization for Standardization/ International Electrotechnical Commission (ISO/IEC) is the most acceptable standard on information security, and the ISO 27005 deals specially with information risks. [33]

We continue our discussion on the security by following the diagram shown in Figure 6.6.

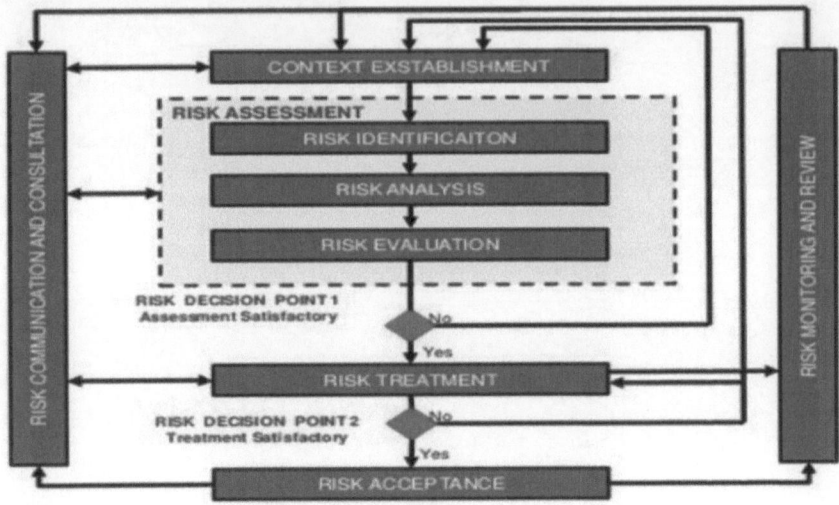

Figure 6.6 The information security risk management process. [34]

We discuss Figure 6.7 in detail next, both in general and with a focus on AV.

6.5.3 Some Terms of IT Security Risks

Several terms are very relevant to both technical and non-technical staff in the organization in order to implement ISRM properly: [37]

- **Threat:** Any probable danger that may happen in future.
- **Vulnerability:** Any weakness that can be exploited by any means.
- **Breach:** When a threat becomes successful through vulnerabilities.
- **Risk:** When a loss is considered low, high, or medium by a breach.
- **Relative risk:** This type of risk signifies multiple risks related to each other.
- **Temporal risk:** This type of risk can be protected by adequate measures.
- **Residual risk:** This type of risk refers to risks that are still present after measures are taken.
- **Compromise:** It happens when vulnerabilities are not considered.
- **Mitigation:** Actions taken to protect activities against different risk elements.
- **Countermeasure:** It refers to threats that are immitigable, which is slightly different from mitigation [38].

6.5.4 Risk Calculation Models

There are several models to calculate risk from different aspects:

- **CRAMM (CTCA [Central Communication and Telecommunication Agency] Risk Analysis and Management Method):** A risk analysis method that involves assets, threats, vulnerabilities associated

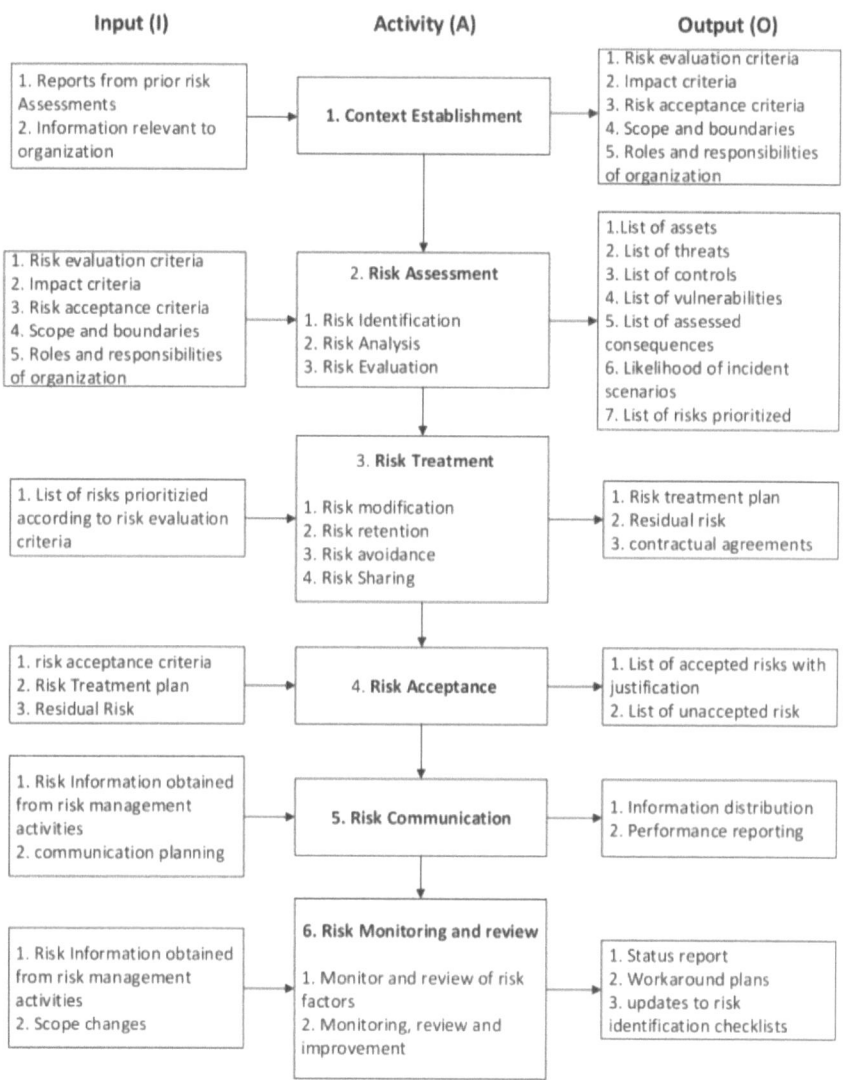

Input (I)   Activity (A)   Output (O)

Input	Activity	Output
1. Reports from prior risk Assessments 2. Information relevant to organization	**1. Context Establishment**	1. Risk evaluation criteria 2. Impact criteria 3. Risk acceptance criteria 4. Scope and boundaries 5. Roles and responsibilities of organization
1. Risk evaluation criteria 2. Impact criteria 3. Risk acceptance criteria 4. Scope and boundaries 5. Roles and responsibilities of organization	**2. Risk Assessment** 1. Risk Identification 2. Risk Analysis 3. Risk Evaluation	1.List of assets 2. List of threats 3. List of controls 4. List of vulnerabilities 5. List of assessed consequences 6. Likelihood of incident scenarios 7. List of risks prioritized
1. List of risks prioritizied according to risk evaluation criteria	**3. Risk Treatment** 1. Risk modification 2. Risk retention 3. Risk avoidance 4. Risk Sharing	1. Risk treatment plan 2. Residual risk 3. contractual agreements
1. risk acceptance criteria 2. Risk Treatment plan 3. Residual Risk	**4. Risk Acceptance**	1. List of accepted risks with justification 2. List of unaccepted risk
1. Risk Information obtained from risk management activities 2. communication planning	**5. Risk Communication**	1. Information distribution 2. Performance reporting
1. Risk Information obtained from risk management activities 2. Scope changes	**6. Risk Monitoring and review** 1. Monitor and review of risk factors 2. Monitoring, review and improvement	1. Status report 2. Workaround plans 3. updates to risk identification checklists

Figure 6.7 ISO27005 standard based on the input and output of each activity [35, 36].

with risks, countermeasures, implementations, and audits (see Figure 6.8).

- **DREAD (Damage, Reproducibility, Exploitability, Affected users, Discoverability):** Microsoft proposes practicing a straightforward DREAD model (see Figure 6.9).
- **STRIDE (Spoofing, Tampering, Repudiation, Information disclosure, Denial of service, Elevation of privilege):** This model is used along with the DREAD model for live systems on the internet (see Figure 6.10).

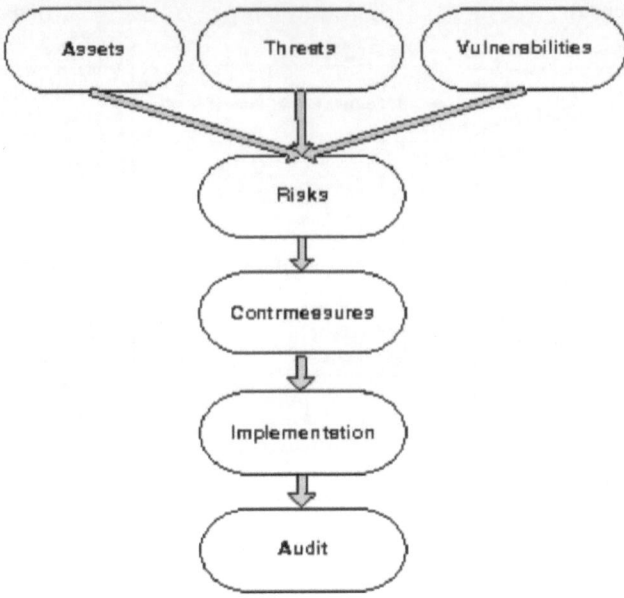

Figure 6.8 CRAMM risk assessment method [39].

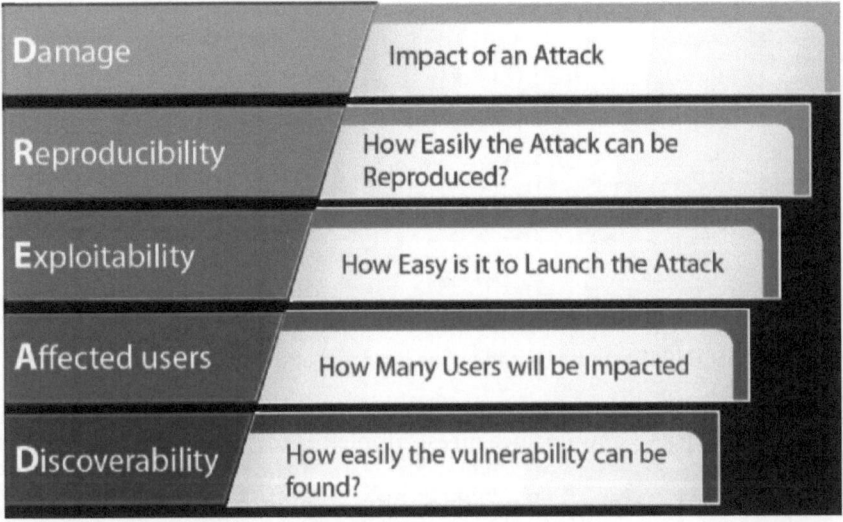

Figure 6.9 DREAD model, threat modeling, and the International Council of Ecommerce
          Consultants (EC-Council) [40].

STRIDE Threat Model

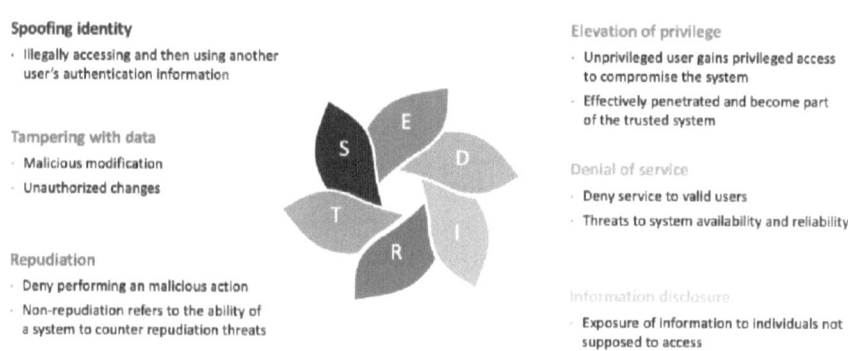

Spoofing identity
- Illegally accessing and then using another user's authentication information

Tampering with data
- Malicious modification
- Unauthorized changes

Repudiation
- Deny performing an malicious action
- Non-repudiation refers to the ability of a system to counter repudiation threats

Elevation of privilege
- Unprivileged user gains privileged access to compromise the system
- Effectively penetrated and become part of the trusted system

Denial of service
- Deny service to valid users
- Threats to system availability and reliability

Information disclosure
- Exposure of information to individuals not supposed to access

Figure 6.10 Stride model [41].

- **FRAP (Facilitated Risk Analysis Process):** This model mostly deals with qualitative aspects [42].
- **OCTAVE (Operationally Critical Threat, Asset, and Vulnerability Evaluation) Allegro:** This model was developed for strategic planning (see Figure 6.11) [43].
- **Spanning Tree Analysis (STA):** This is mainly used to analyze threats [45].

6.5.4.1 Context establishment

To establish the overall context, it is important to understand both the internal and external contexts.

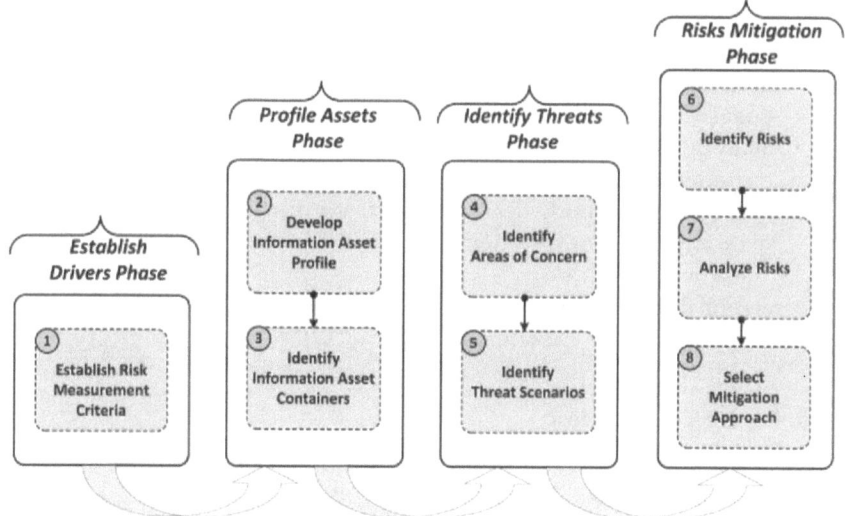

Figure 6.11 Octave Allegro [44].

The external context includes social, cultural, environmental (for example, natural calamities and climate issues), political, legal, financial, technological, security, and economic factors, as well as external stakeholders' parameters, and so on.

The internal context includes strategic objectives, values, standards, resources available, capacities, business processes, organizational culture, internal stakeholders' factors, and so on. Stakeholders' factors refers to their perceptions, relationships, and expectations of an organization. Our defined system would be evaluated on the basis of following context: Governance, Business Continuity and Succession Planning, Business, Financial, Regulatory, Technology, Human resources, Stakeholder (Figure 6.12).

6.5.4.2 Risk identification

We have to keep in mind that for effective system design, it is urgent to design it in such a way that every possible risk would be discussed in order to eliminate any potential incidents. This is why we must use a comprehensive risk analysis approach that begins with subsequent phases of context establishment. An example of risk identification is shown in Table 6.2 [47].

6.5.4.3 Risk analysis

Risk analysis is associated with risk exposure, and likelihood and severity of risks with proper screening that concentrates on source, history, statistics, Who/How (W/H) information, and so on to gain an understanding of the background, vulnerability and credibility.

- **Quantitative methods** are followed by modelling, statistical analysis, decision trees, event tree analysis, probability analysis, and consequence analysis.
- **Semi-quantitative methods** focus on a quantitative score to a qualitative approach; for example, high risk, medium risk, or low risk.
- **Qualitative methods** deal with brainstorming, specialist knowledge, questionnaires, or intuition.

The relationship among consequences, likelihood, and level of risk is
Risk = consequence * likelihood [48].

6.5.4.4 Risk evaluation

In this step, we have to consider the organizations capacity, as it varies from one organization to another. On top that, it also depends on factors such as: [49]

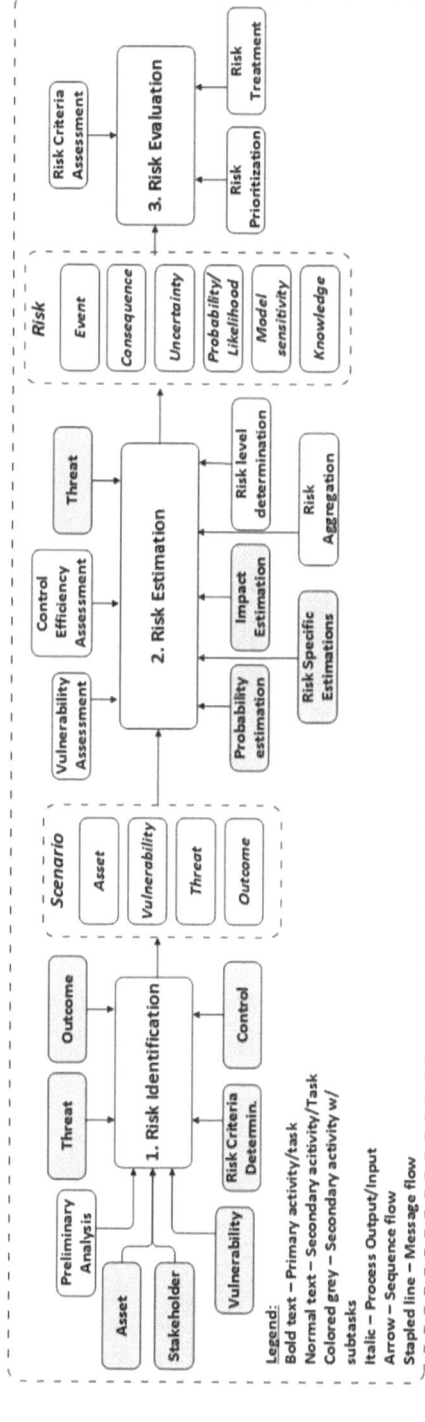

Figure 6.12 The generic output of the risk evaluation is prioritized risks; source [46].

Table 6.2 Risk analysis with scoring and ranking

Serial	Asset	Domain	Threat	Vulnerability	Loss of CIA	Severity 1-5	Likelihood 1-5	Risk 1-25	Additional Risk Control Measures	Severity 1-5	Likelihood 1-5	Residual 1-25	Remarks
1	Web Server	Physical	Being Stolen	Lack of Physical Security	C+A	5	2	10	Improve Physical Security	5	1	5	
2	Web Server	Physical	Outdated or Non-functional	Lack of Attention or Supervision	A	2	3	6	Improve Supervision	2	1	2	
3	Web Server	Physical	OS/Server System Fail	Irregular Monitoring on OS	A	2	2	4	Improve Monitoring	2	1	2	
4	Web Server	Physical	Harddisk Crash	Lack Update of Hardwares	A	2	3	6	More Attention	2	1	2	
5	Web Server	Virtual	Storage DOS	filling resources	A	2	3	6	Set disk quota's for users	1	1	1	
6	Web Server	Cloud	Earthquake or War Bombing	Out of Control	A	1	1	1	Selection of Safe Zones	1	1	1	
7	Access Point	Router/Switch	Attack Surface	DoS	C+I+A	5	3	15	VAPT and Proxy Server using	5	1	5	
8	Access Point	Physical	Tamper	Raspberry Pi has many open port	C+A+I	4	2	8	Use GPIO with Light Detection to self destruct	1	1	1	
9	Access Point	Router/Switch	Outdated Firmware	Loosing access control	C+A+I	5	2	10	Regular Firmware Upadating by experts	5	1	5	
10	Access Point	Virtual	Weak Passwords	unauthorised access	C	4	4	16	Change password, implement policy	4	1	4	
11	Web Site	Virtual	SQL Injection	Loss of data	C+I+A	5	2	10	Adequate filtering in database	5	1	5	
12	Website	Virtual	XSS Scripting	Loss of data	C	5	2	10	Securing browser and certificates on token	5	1	5	
13	Firewall	Virtual	Inside Attack & Security Patches	Loosing access control	C+I+A	5	2	10	Strong administrative and technical control	5	1	5	
14	DNS	Virtual	Cache Poisoning Attacks	send legitimate requests to malicic	C	3	2	6	Use VPN, HTTPS, Encryption	3	1	3	
15	Proxy Server	Virtual	SSL-based DDoS attacks	latency and audit issues	A	4	2	8	Using strong firewall and continuous monitoring	4	1	4	
16	Workstation Dev	Physical	Physical damage or data leakage	Data loss	C+I+A	5	1	5	Strong physical security system like entry-exit	5	1	5	
17	User Device	Both	Physical damage or virtual control	loss of data and functionality	C+I+A	5	1	5	Awareness and Training	5	1	5	

- Organizational policies
- Goals and objectives
- The attitude of internal and external stakeholders
- The environment and geographic location
- Managerial structures
- Constraints, budget allocations, and resiliency
- Legal limitations
- Approximate maintenance costs
- Possible fines by regulatory authorities
- Ethical, humanitarian, or moral practices
- Reputation and brand value

6.5.4.5 Risk treatment

Risk treatment consists of avoiding, reducing, sharing, transfer, or accepting risks. It might differ with risk type, magnitude, and other relevant factors.

Table 6.2 shows a typical concept of risk assessment for an information system where risk management approach has been performed according to ISO 27005. [50]

6.6 RISK MANAGEMENT FOR AV

Given it improves transportation systems in terms of physical security, user friendliness, environmental effects, and traffic congestion, the AV is a desirable addition to science, engineering, and technology. Continuous advancements in sensor, computer, and tele-communication improved its performance, though it has some new challenges, such as cybersecurity and interactions with other non-AVs. The sections that follow address several risks associated with the behavior of AV in a mixed traffic environment with regard to cybersecurity aspects outlined in published scientific articles and other relevant data sources. Low-power lasers and pulse generators can build up a system to make AV sensors or Lidar fool. Hacker can take controls of the brakes, components, and other elements remotely. So it is crucial to examine all possible risks associated with AVs to prevent any future accidents [51].

The purpose of a risk analysis is to ensure a safer system by protecting it from inherent threats of failure from existing vulnerabilities. A fault tree analysis is used to identify chronological and sequential catastrophes within a system, much like designing aircraft processes or nuclear power plants, and analyzing of risks of an AV in mixed traffic streams. The difference between the actual outputs and the expected outputs quantifies the failure. The fault tree analysis method can provide the minimum path for a single-component or full-vehicle failure.

A safe and reliable transportation system can be obtained through a comprehensive risk assessment, which is conducted by several researchers for an individual subsystem of AV. Surveyed motion prediction and relevant risk assessment methods usually look for collision predictions, which are costly in case of improving efficiency to consider physical parameters of AV, weather conditions, road-surface conditions, speeds approaching an intersection, real trajectories, the fatigue level of the driver, data from vehicle sensors, the lane departure warning and driving assistance subsystem, and pedestrian behavior in different traffic scenarios. The Bayesian belief network approach method only considers optical systems like the camera and Lidar.

Figure 6.13 is an example of a risk assessment of an AV, which includes risk identification, risk estimation, and an evaluation of fault tree model.

Table 6.3 highlights all the basic events related to AV along with descriptions. It also demonstrates which methods and experiments can be applied for analysis to calculate probabilities of failure. Non-AVs crashes usually result from reckless driving, fatigue, hardware failure, and distractions,

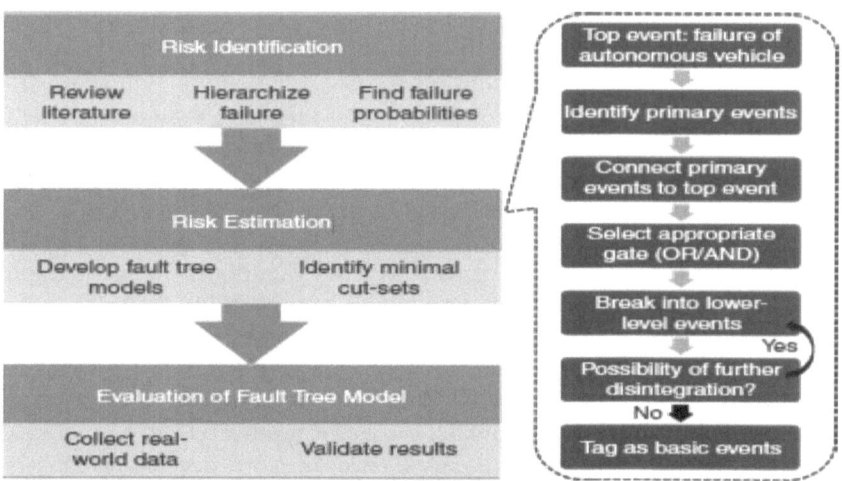

Figure 6.13 The general research pattern for risk analysis [51].

Table 6.3 Failure probabilities, experiment types, methods, and events of AV components [52]

Basic Event	Description	Methods	Experiment Type	Failure Probability (%)
Lidar failure	Laser malfunction, mirror motor malfunction, position encoder failure, overvoltage, short-circuit, optical receiver damages	Bayesian belief network	Simulation	10.0000 (48)
Radar failure	Detection curves drawn with respect to signal and noise ratios	Chi-square distribution	Mathematical modeling	2.0000 (51)
Camera failure	Foreign particles, shock wave, over-voltage, short-circuit, vibration from rough terrain, etc.	Bayesian belief network	Simulation	4.9500 (48)
Software failure	System had to generate outputs from array definition language statements	Extended Markov Bayesian network	Experiment (3,000 runs)	1.0000 (52)
Wheel encoder failure	Encoder feedback unable to be transferred, which can cause loss of synchronization of motor stator and rotor positions	Kalman filter	Experiment	4.0000 (53)
GPS failure	Real-life tests performed with high-sensitivity GPS in different signal environments (static and dynamic) for more than 14 h	Least squares	Experiment (at 4 locations)	0.9250 (54)
Database service failure	Using new empirical approach, connectivity and operability data of a server system were collected.	Generic quorum-system evaluator	Experiment (for 191 days)	3.8600 (55)
Communication failure	Wi-Fi: Periodic transmission of 1,000-byte frames (average conditional probability of success after previous success considered)	In IEEE 802.11b network	Experiment (with 10 vehicles)	5.1250 (56)
	LTE: Network unavailability during location update in mobility was considered here	Application of CAP theorem	Experiment	5.8800 (57)
Integrated platform failure	A two-state model with failure rates was developed to estimate the computer system availability.	Markov chain model	Mathematical modeling	2.0000 (58)
Human command error	Three data sets of over 115 months from NASA were analyzed and then validated by three methods (THERP, CREAM, and NARA) to facilitate NASA risk assessment.	Human reliability analysis	Experiment (from December 1998 to June 2008)	0.0530 (59)
System failed to detect human command	System unable to detect the accurate acoustic command; driver inputs the wrong command, and system unable to detect wrong commands	Artificial neural networks on clean speech	Experiments (37 subjects: 185 recording)	1.4000 (60)

which are considered here. Cyclists responsible for crashes with AV are also included here. Accidents including pedestrians and construction zones are included with AV accidents. Weather-related incidents include fog, mist, sleet, snow, dust, rain, severe crosswind, thunderstorm, and smoke. Road conditions such as improper lane marking, holes in the road, ice formations, fallen trees, and pavement conditions may cause crashes as well.

A "fixed probability" statistical model can be used to analyze risks on the failure of vehicular components of AV (see Figure 6.14) and Isograph FaultTree+ software models the distribution of probability for basic event failure (see Figure 6.15). The fault tree for the infrastructure component failures for the infrastructure components appear in Figure 6.16, which is based on the fault tree. Combined Fault Tree is obtained by bringing the experimental estimation and simulation modeling together to form a risk hierarchy for upholding risk rankings in terms of score, developing cut-sets of the main event, and it can assist in improving the safety performance of the AV.

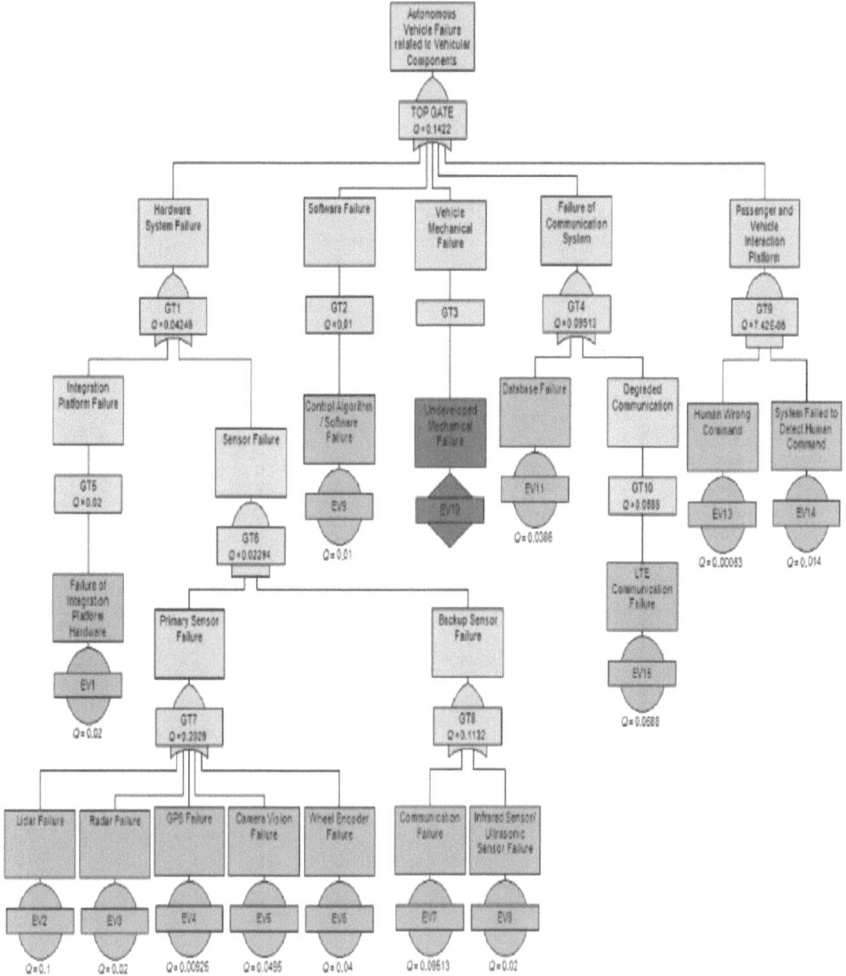

Figure 6.14 A fault tree analysis considering failures related to vehicular components (*Q* = probability value either inputted into the fault tree or calculated by fault tree analysis) [52].

AVs can contribute many ways to ensuring a safe, secure, and sustainable transportation system for the next generation, and proper risk analysis can aid in reducing losses in real life before any large-scale deployment is deployed. In-depth fault identification by tree analysis might help developers and researchers to reduce probable risks, before an AV is implemented, in a comprehensive approach using continuous innovations in computing and communication technologies, though some limitations for reckless human drivers in a mixed traffic system exist.

Nowadays, researchers are very keen to study cybersecurity with AVs where they have categorized threats as high, medium, and low based on the

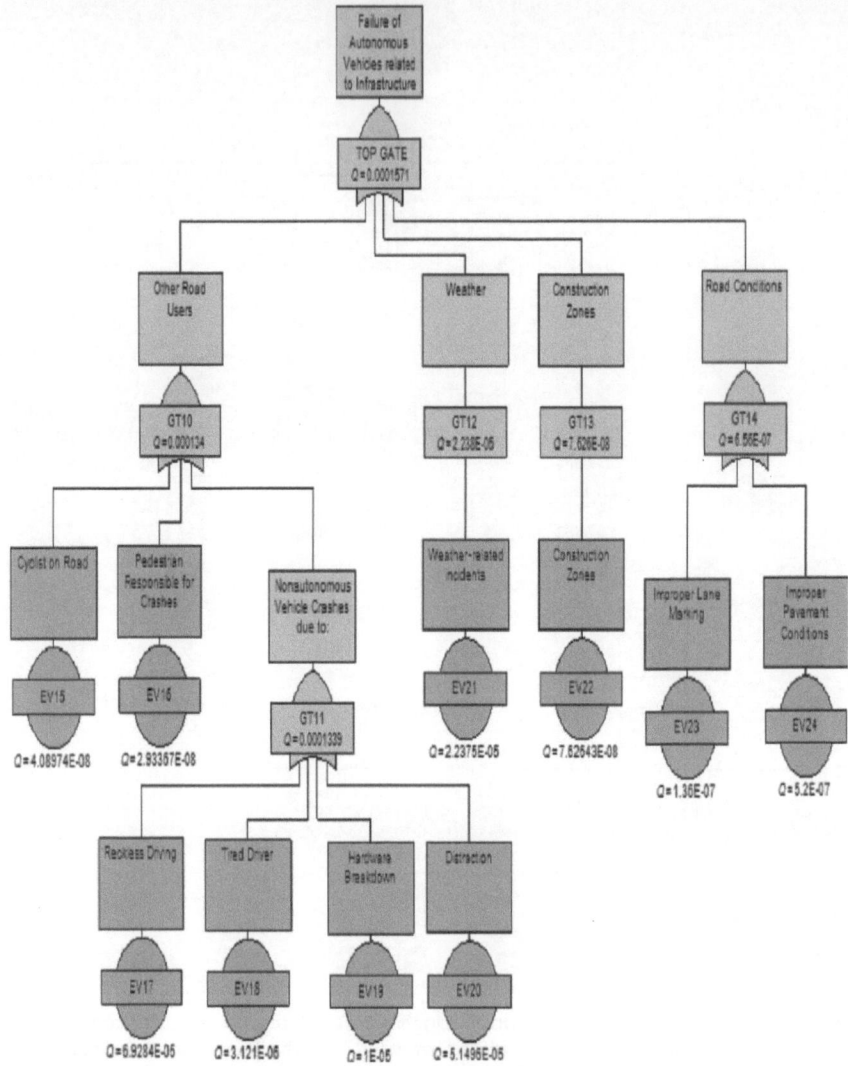

Figure 6.15 A fault tree analysis considering failures related to transportation infrastructure components [53].

feasibility, probability, frequency, and success rate of every attack. In order to create anti-attack techniques, researchers have gone through attack and defense scenarios; examples of the steps taken are discussed a little later in this chapter.

Graphs to realize the computer networks' security as a powerful tool for analyzing the risks for vulnerable components in AVs. These graphs help visualize all countermeasures against each risk [54].

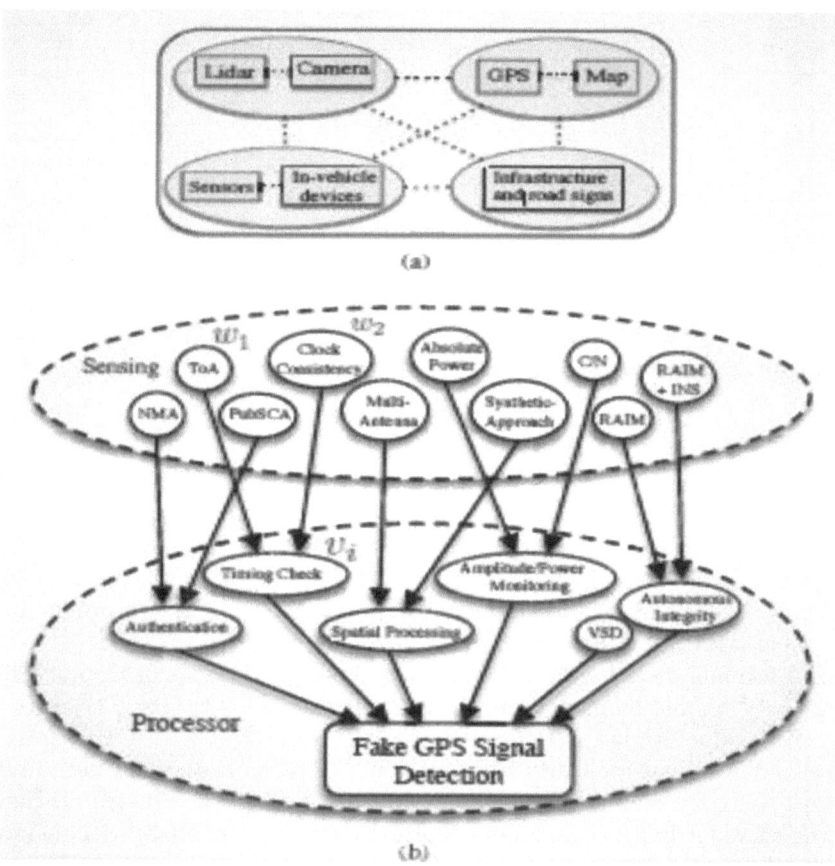

Figure 6.16 (a) Security monitoring unit, and (b) graphical model for secured GPS component in AV [56].

The proposed plain security model can monitor vulnerable components using security nodes. E-Safety Vehicle Intrusion Protected Applications (EVITA) represents countermeasures in a graph with all security states.

EVITA and Common Vulnerability Scoring System (CVSS) are two widely used methodologies for conducting a risk assessment of an AV based on likelihood, impact, and severity. Safety, privacy of drivers, operational performance, and financial losses are four important parameters for calculating the risks of a vehicle and can be categorized as none, low, medium, and high [55].

The Bayesian defense graph studies the cyber-related risks and threats of AVs; for example, EVITA has been applied to a framework for GPS signals. However, the likelihood of threats from existing vulnerabilities can be declined to 0:01 percent using anti-spoofing techniques that can mobilize future work [57].

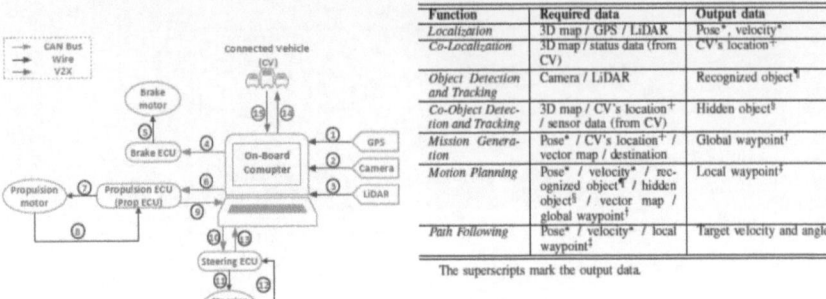

Function	Required data	Output data
Localization	3D map / GPS / LiDAR	Pose*, velocity*
Co-Localization	3D map / status data (from CV)	CV's location+
Object Detection and Tracking	Camera / LiDAR	Recognized object¶
Co-Object Detection and Tracking	3D map / CV's location+ / sensor data (from CV)	Hidden object§
Mission Generation	Pose* / CV's location+ / vector map / destination	Global waypoint†
Motion Planning	Pose* / velocity* / recognized object¶ / hidden object§ / vector map / global waypoint†	Local waypoint‡
Path Following	Pose* / velocity* / local waypoint‡	Target velocity and angle

The superscripts mark the output data.

Figure 6.17 Data acquisition strategy for VeRA model.

Vehicles Risk Analysis (VeRA) is an appropriate method for analyzing the different risks of a single AV and Connected AV (CAV) considering multiple factors (standard SAEJ3061). It is comparatively an easier approach that discusses three important parameters simultaneously, and we will also look at a comparison of VeRA with some other models (Figure 6.17). The knowledge (K) as well as equipment (E) determines the attack probability (P) has (see Table 6.4).

A formula can be obtained by the combination of attack probability (P), severity (S), and human control (H) as follows:

Risk value, R= H + S*P

Risk Analysis for Cooperative Engines (RACE) is another method of risk analysis that is specially developed for CAV; it is more efficient than EVITA but it requires much effort keeping controllability constant (Tables 6.5–6.7) [58].

Attack probability, severity, and a vehicle's automation level form the US2 method for AVs to determine safety, security, and attacks defined by the SAE standard J3016, excluding CAVs that extend to the Cyber Security Risk Level (CSRL) for CAVs. The CSRL method is slow and it does not take automation and human control into account [59].

Table 6.4 VeRA risk value calculation using formula.

VeRA RISK CLASSIFICATION MATRIX, ACCORDING TO H, S AND P.

ATTACK PROBABILITY P CLASSIFICATION, AND CORRESPONDING $K + E$.

Attack probability P	Description	Corresponding $(K+E)$
1	Low	$K + E > 2$
2	Median	$1 < K + E \leq 2$
3	High	$K + E \leq 1$

Human Control H	Severity S	Attack probability P 1	2	3
1	1	2	3	4
	2	3	5	7
	3	4	7	10
2	1	3	4	5
	2	4	6	8
	3	5	8	11
3	1	4	5	6
	2	5	7	9
	3	6	9	12

Table 6.5 VeRA severity scale

Severity S	Safety S_s	Privacy S_p	Financial (in €) S_f	Operational S_o
0	No injuries	No unauthorized access to data	$0 < loss < 100$	No impact on performance
1	Light injuries	Access to anonymous data	$100 < loss < 1000$	Impact not detected by driver
2	Severe injuries, with survival	Identification of car or driver	$1000 < loss < 10000$	driver aware of performance degradation
3	Life threatening, possible death	Driver or car tracking	$loss > 10000$	Significant impact on performance

The international Society of Automotive Engineers (SAE) has found six levels of automation, which are outlined in Table 6.8 [60].

The ISO 26262 process is currently widely used for risk analysis of AVs, which consists of the following, and is shown in Table 6.9:

- Item definition,
- Hazard and risk analysis
- ASIL assignment
- Determination of safety goals [62].

6.7 ISSUES

The Internet of Autonomous Vehicles (IoAV) platform upholds an improved transportation system, including better fuel economy, the latest safety standards, minimum road accidents, and a more comfortable journey experience. However, it is associated with multiple issues, such as device design, data management, social, ethical, professional, legal, and security issues.

Table 6.6 Relationship among different models and inter-connections

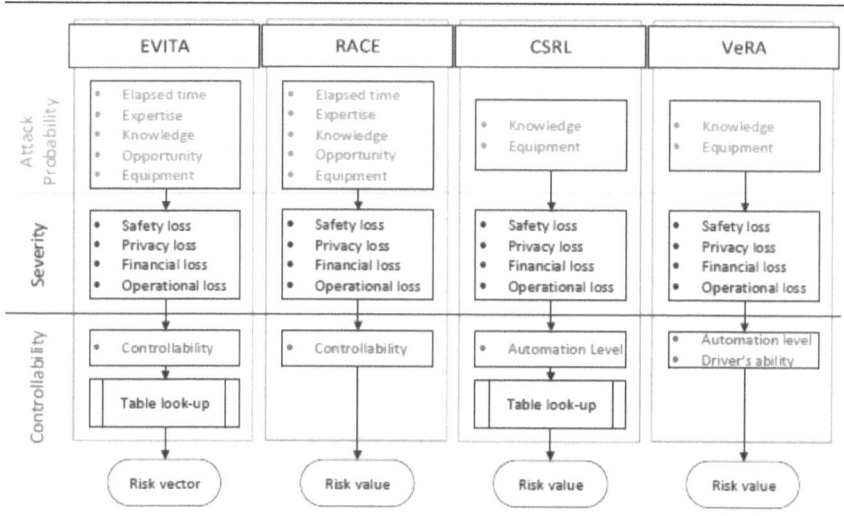

Table 6.7 Description of different types of attacks

Detailed attack	Description
Alteration	Modify configuration files
DoS	Make network link unavailable by flooding the network interface controller with bad traffic packets
Eavesdropping	Capture secretly and analyse the exchanged messages
Sensor Jamming	Interfere with sensor antennas by saturating their receivers with noise and false formation, e.g., GPS jamming, LiDAR jamming
Malware	Run a malware on an equipment (e.g., ECU, external sensor), either by flashing it directly into the ECU memory or using the firmware update process
Message injection	Inject self-created information or command on a network link (Ethernet, CAN or wireless)
Message suppression	Delete messages on a network link
Replay	Replay of past sniffed traffic
Sensor spoofing	Manipulate sensor to fake the captured and returned information. This attack includes rogue GPS attack where we generate radio signals to simulate a GPS signal to make the receiver use fake position information (i.e., GPS Spoofing)
Sybil	A malicious vehicle transmits the various messages with multiple fake or stolen source identities to other vehicles
Falsified entity attack	The attacker passes information to the connected vehicle through a valid network identifier

6.7.1 Ethical issues

The ethical concepts of highly autonomous vehicles (HAVs) are addressed by manufacturers, consumers, government, and stakeholders, and they are critical concepts with many dilemmas. It can be handled through policies, but still creates conflicts among the following:

- An individual's interest
- The community's interest
- Protecting people from harm
- Providing people with healthy lives

A specific example of this dilemma is shown in the trolley problem [65].

The common ethical stand for all information systems is providing safety and reliability among all types of information and data handling, though it varies from system to system, like open or close networks, local

Table 6.8 Levels of autonomy by SAE [61]

SAE level	Name	Narrative Definition	Execution of Steering and Acceleration/ Deceleration	Monitoring of Driving Environment	Fallback Performance of Dynamic Driving Task	System Capability (Driving Modes)
Human driver monitors the driving environment						
0	No Automation	the full-time performance by the *human driver* of all aspects of the *dynamic driving task*, even when enhanced by warning or intervention systems	Human driver	Human driver	Human driver	n/a
1	Driver Assistance	the *driving mode*-specific execution by a driver assistance system of either steering or acceleration/deceleration using information about the driving environment and with the expectation that the *human driver* perform all remaining aspects of the *dynamic driving task*	Human driver and system	Human driver	Human driver	Some driving modes
2	Partial Automation	the *driving mode*-specific execution by one or more driver assistance systems of both steering and acceleration/deceleration using information about the driving environment and with the expectation that the *human driver* perform all remaining aspects of the *dynamic driving task*	System	Human driver	Human driver	Some driving modes
Automated driving system ("system") monitors the driving environment						
3	Conditional Automation	the *driving mode*-specific performance by an *automated driving system* of all aspects of the dynamic driving task with the expectation that the *human driver* will respond appropriately to a *request to intervene*	System	System	Human driver	Some driving modes
4	High Automation	the *driving mode*-specific performance by an automated driving system of all aspects of the *dynamic driving task*, even if a *human driver* does not respond appropriately to a *request to intervene*	System	System	System	Some driving modes
5	Full Automation	the full-time performance by an *automated driving system* of all aspects of the *dynamic driving task* under all roadway and environmental conditions that can be managed by a *human driver*	System	System	System	All driving modes

Table 6.9 Application of ISO 26262 with STPA for engine systems [63, 64]

Item	Malfunction	Driving Situation	Risk Scenario	Hazard
Engine System (S7)	f1: Accelerates more than intended	Urban driving environments with medium speeds between 25 mph and 40 mph	Traffic enroute or presence of obstacles	Collision possible with infrastructure or pedestrians
	f2: Missing feedback from sensors regarding vehicle state	Freeway at high speeds (> 40 mph)	Heavy obstacle present at $< 300m$	Collision possible with traffic or infrastructure
	f3: Erroneous throttle valve signal	Freeway at high speeds (> 40 mph)	Minimum preferred distance to obstacle exceeded	Collision possible with traffic or infrastructure

Safety Goal
SG1: Avoid unintended increase in engine torque in Urban areas beyond t1 time period
SG2: Avoid unintended decrease in engine torque in Urban areas beyond t2 time period
SG3: Avoid unintended acceleration in urban areas beyond t1 time period
SG4: Avoid unintended deceleration beyond t2 time period
SG5: Ensure integrity in sensor feedback

area networks (LANs), internet service providers (ISPs), backbone sites, and so on.

A system administrator would definitely focus on all possible threats to provide privacy, liability, and patent and copyright laws. The system has to ensure confidentiality, trade secrets, fraud and misuse cases, and have a decision-making capacity based on ethics and responsibility [66].

6.7.2 Environmental issues

Environmental issues relate mainly to energy consumption, car emissions, expired parts reclining, and so on. Thus, the AV engineer, designer, and researcher must consider ways to ensure low rates of fuel consumption, an enhanced use of renewable energy, new technologies to reduce carbon emissions, high-quality road designs to decrease frictional wear, and the shortest routes of travel (see Figure 6.18).

6.7.3 Legal issues

Legal concerns vary among locality, state, region, country, policy, and government. However, AV must deal with two different aspects of law, one is road transport law and another is information system domain law. The overall system relates to roads, transport, information, information systems, data, finance, cybersecurity, computer systems, people, and so on. Thus, it is a multi-disciplinary sector that mostly deals with public data. Several legal frameworks are available based socio-economic and geo-political circumstances like General Data Protection Rule (GDPR), Data Protection

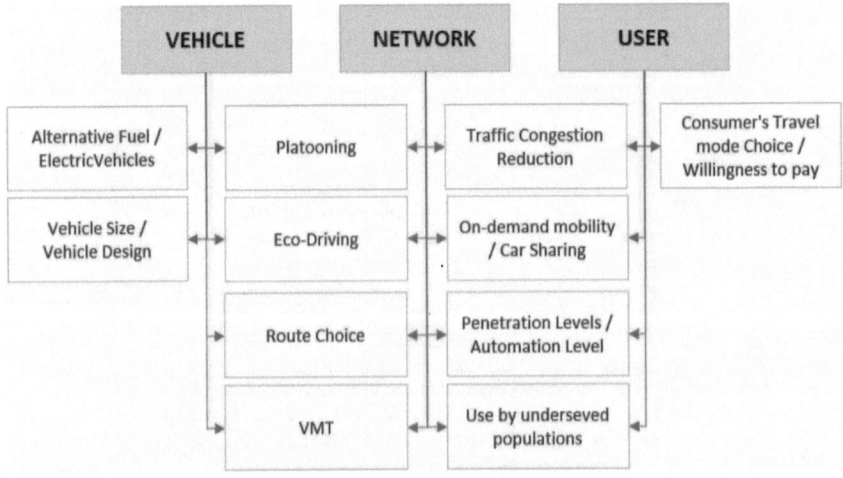

Figure 6.18 Key elements for environmental concern [67].

Act (DPA), BSC, Institute of Electrical and Electronics Engineers (IEEE), BCS, and IET, and so on. We have to abide by all these issues to operate the system as a white one. [68–70].

- Traffic acts
- Vehicle acts
- GDPR
- Data Protection Act
- Road transport law

6.7.4 Professional issues

AV professionals must maintain moral and ethical values, along with existing laws and regulations for data security, automobile design, and road system. It includes all people handling AV operations to some extent, such as the drivers, robots, companies, engineers, data administrators, and so on. All personnel must be libel for any incidents to ensure accountability [71].

6.7.5 Social issues

One of the social dilemmas with AV systems is exemplified by this hypothetical scenario: ten pedestrians appear in the vehicle's path, and the vehicle is not able to break early enough, should the algorithm in the AV save the ten pedestrians or the passenger? One study by Bonnefon et al. (2016), investigated human morality when it comes to those two algorithms, the utilitarian approach that chooses to save a greater number of lives, and the non-utilitarian approach that chooses to save the passenger (*passenger-protective*). Given this dilemma, participants were required to rate which action they thought was more moral. The result of this survey was that the participants found the utilitarian algorithm to be the moral choice. However, in another survey was taken where the participants were asked about the likelihood of purchasing the vehicle of each algorithm, the study found that people rated the vehicle with the non-utilitarian algorithm (passenger-protected) one that they were more likely to purchase. A social dilemma arises due to these surprising results. There is the temptation to act according to self-interest, which usually leads to the worst results for everyone involved [28].

The research created uncertainty in the number of participants. It is possible that the responses were different for each question because of participants had limited access to moral utilitarian information. One example would be that participants envisioned themselves as the driver, although the driver would also be, at some point, the pedestrian. To eliminate any conflict in the participants' responses, complete descriptions of the utilitarian approach should be provided, including the utilitarian tasks and their consequences.

Psychological uncertainty is found in different utilitarian choices for scenarios that are morally sensitive. The uncertainty can be reduced using comprehensive moral questions and tasks. When it came to Bonnefon's study, the utilitarian information was limited. In the study, participants were asked to envision themselves as the passenger, thereby giving the only one side of the situation. The proposal was made that asking participants to envision themselves as the pedestrian too would help solve the issue of uncertainty. This way participants have both views of the situation, the passenger's view and the pedestrian's view [29].

6.8 CONCLUSION

The fourth industrial revolution demands automation where AVs are a key element. We have covered primary aspects are in this chapter. Smart transportation does not literally mean AV, but it is synonymous to AV in depth realization. Urban planners must work to build smarter cities, with AV in mind. Additionally, the challenges related to the maintenance of AVs, such as road design, fuel management, security issues, and so on, must be considered. This chapter covered security concerns and the issues it raises. Readers will have gained a passion for AVs while going through this chapter. It provides a good start on insights with AV. However, given it is merely a chapter, it is a springboard for readers to seek out more.

REFERENCES

[1] Schwarz, C., Thomas, G., Nelson, K., McCrary, M., Sclarmann, N., and Powell M. *Towards autonomous vehicles*. No. MATC-UI: 117. Mid-America Transportation Center, 2013.

[2] Fagnant, D. J., and Kockelman, K. (2015) "Preparing a nation for autonomous vehicles: opportunities, barriers and policy recommendations." *Transportation Research Part A: Policy and Practice*, 77, pp. 167–181.

[3] Schwarting, W., Alonso-Mora, J., and Rus, D. (2018) "Planning and decision-making for autonomous vehicles." *Annual Review of Control, Robotics, and Autonomous Systems*.

[4] Kato, S., Takeuchi, E., Ishiguro, Y., Ninomiya, Y., Takeda, K., and T. Hamada, "An Open Approach to Autonomous Vehicles." *IEEE Micro*, 35(6), pp. 60–68, Nov.-Dec. 2015, doi: 10.1109/MM.2015.133.

[5] Kocić, J., Jovičić, N., and Drndarević, V. "Sensors and Sensor Fusion in Autonomous Vehicles," 2018 26th Telecommunications Forum (TELFOR), Belgrade, 2018, pp. 420–425, doi: 10.1109/TELFOR.2018.8612054.

[6] Zehang Sun, Bebis, G., and Miller, R., "On-road vehicle detection: a review." *IEEE Transactions on Pattern Analysis and Machine Intelligence*, 28(5), pp. 694–711, May 2006, doi: 10.1109/TPAMI.2006.104.

[7] Patole, S. M, Torlak, M., Wang, D., and Ali, M., "Automotive radars: A review of signal processing techniques." *IEEE Signal Processing Magazine*, 34(2), pp. 22–35, March 2017, doi: 10.1109/MSP.2016.2628914.

[8] Khader M. and Cherian D., "An Introduction to Automotive LIDAR," Texas Instruments, May. 2020. Accessed on: Dec. 27, 2020. [Online]. Available: https://www.ti.com/lit/wp/slyy150a/slyy150a.pdf?ts=1609169652732&ref_url=https%253A%252F%252Fwww.google.com%252F

[9] Thakur, R., "Scanning LIDAR in Advanced Driver Assistance Systems and Beyond: Building a road map for next-generation LIDAR technology." *IEEE Consumer Electronics Magazine*, 5(3), pp. 48–54, July 2016, doi: 10.1109/MCE.2016.2556878.

[10] Levinson, J. *et al.*, "Towards fully autonomous driving: Systems and algorithms," 2011 IEEE Intelligent Vehicles Symposium (IV), Baden-Baden, 2011, pp. 163–168, doi: 10.1109/IVS.2011.5940562.

[11] Durrant-Whyte, H., Rye, D., and Nebot, E. (1996) "Localization of autonomous guided vehicles." *Robotics Research*. London: Springer. 613–625.

[12] Levinson, J., Montemerlo, M., and Thrun, S. (2007) "Map-Based Precision Vehicle Localization in Urban Environments." *Robotics Science and Systems*.

[13] Teichman, A. and Thrun, S. (2011) "Practical object recognition in autonomous driving and beyond," *Advanced Robotics and its Social Impacts*, CA: Half-Moon Bay, pp. 35–38, doi: 10.1109/ARSO.2011.6301978.

[14] Teichman, A., Levinson, J., and Thrun, S., "Towards 3D object recognition via classification of arbitrary object tracks," 2011 IEEE International Conference on Robotics and Automation, Shanghai, 2011, pp. 4034–4041, doi: 10.1109/ICRA.2011.5979636.

[15] Goldberg, K. and Kehoe, B. "Cloud robotics and automation: A survey of related work." *EECS Department, University of California, Berkeley, Tech. Rep. UCB/EECS-2013-5* (2013).

[16] Hunziker, D., Gajamohan, M., Waibel, M., and D'Andrea, R., "Rapyuta: The roboearth cloud engine," *Robotics and Automation (ICRA) 2013 IEEE International Conference*, pp. 438–444, 2013.

[17] Lee, J., Wang, J., Crandall, D., Šabanović, S., and Fox, G., "Real-Time, Cloud-Based Object Detection for Unmanned Aerial Vehicles," 2017 First IEEE International Conference on Robotic Computing (IRC), Taichung, 2017, pp. 36–43, doi: 10.1109/IRC.2017.77.

[18] Philip, B. V., Alpcan, T., Jin, J., and Palaniswami, M., "Distributed real-time IoT for autonomous vehicles." *IEEE Transactions on Industrial Informatics*, 15(2), pp. 1131–1140, Feb. 2019, doi: 10.1109/TII.2018.2877217.

[19] Minovski, D., Åhlund, C., and Mitra, K. (May 2020) "Modeling Quality of IoT Experience in Autonomous Vehicles." *IEEE Internet of Things Journal*, 7(5), pp. 3833–3849, doi: 10.1109/JIOT.2020.2975418.

[20] Dilley, J., Poelstra, A., Wilkins J., Piekarska, M., Gorlick, B., and Friedenbach, M., "Strong federations: An interoperable blockchain solution to centralized third party risks", *arXiv preprint arXiv:1612.05491*, 2016.

[21] Guo, H., Meamari, E., and Shen, C., "Blockchain-inspired Event Recording System for Autonomous Vehicles," 2018 1st IEEE International Conference on Hot Information-Centric Networking (HotICN), Shenzhen, 2018, pp. 218–222, doi: 10.1109/HOTICN.2018.8606016.

[22] *Umsl.edu*, 2020. [Online]. Available: http://www.umsl.edu/˜siegelj/information_theory/projects/HashingFunctionsInCryptography.html. [Accessed: 30- Dec-2020].

[23] Kenney, J. B. (July 2011) "Dedicated Short-Range Communications (DSRC) standards in the United States." *Proceedings of the IEEE*, 99(7), pp. 1162–1182, doi: 10.1109/JPROC.2011.2132790.

[24] He, G. and Ma, S. (2002) "A study on the short-term prediction of traffic volume based on wavelet analysis." *Proceedings of IEEE International Conference* on *Intelligent Transportation Systems*, pp. 731–735.

[25] Chen, D. (Aug. 2017) "Research on traffic flow prediction in the big data environment based on the improved RBF neural network." *IEEE Transactions on Industrial Informatics*, 13(4), pp. 2000–2008, doi: 10.1109/TII.2017.2682855.

[26] Karaboga, D., and Bahriye, B. (2008) "On the performance of artificial bee colony (ABC) algorithm." *Applied soft computing*, 8(1), pp. 687–697.

[27] Okdem S., Karaboga, D., and Ozturk, C., "An application of Wireless Sensor Network routing based on Artificial Bee Colony Algorithm," 2011 IEEE Congress of Evolutionary Computation (CEC), New Orleans, LA, 2011, pp. 326–330, doi: 10.1109/CEC.2011.5949636.

[28] Bonnefon, J. F., Shariff, A., and Rahwan, I. (2016). "The social dilemma of autonomous vehicles." *Science*, 352, pp. 1573–1576. doi: 10.1126/science.aaf2654

[29] Martin, R., Kusev, I., Cooke, A. J., Baranova, V., Van Schaik, P., and Kusev, P. (2017) "Commentary: The social dilemma of autonomous vehicles." *Frontiers in Psychology*, 8, p. 808.

[30] www.analyticsinsight.net/top-10-autonomous-vehicle-companies-watch-2020/

[31] www.alliedmarketresearch.com/autonomous-vehicle-market

[32] builtin.com/transportation-tech/self-driving-car-companies

[33] Patiño, S., Solís, E.F., Yoo, S.G., and Arroyo, R., 2018, April. ICT risk management methodology proposal for governmental entities based on ISO/IEC 27005. In *2018 International Conference on eDemocracy & eGovernment (ICEDEG)* (pp. 75–82). IEEE.

[34] Wangen, G., Hallstensen, C., and Snekkenes, E., 2018. "A framework for estimating information security risk assessment method completeness." *International Journal of Information Security*, 17(6), pp. 681–699.

[35] Information technology–security techniques–information security risk management. Iso, 2011.

[36] Agrawal, V., "A Framework for the Information Classification in ISO 27005 Standard," 2017 IEEE 4th International Conference on Cyber Security and Cloud.

[37] Lepofsky, R. (2014) *the manager's guide to web application security: a concise guide to the weaker side of the web*. Berkeley, CA: Apress (The expert's voice in security). doi: 10.1007/978-1-4842-0148-0.

[38] Han, Z. *et al.* (2016) "Risk Assessment of Digital Library Information Security: A Case Study," *The Electronic Library*, 34(3), pp. 471–487. doi: 10.1108/EL-09-2014-0158.

[39] El Fray, I., 2012, September. A comparative study of risk assessment methods, MEHARI & CRAMM with a new formal model of risk assessment (FoMRA) in information systems. In IFIP International Conference on Computer Information Systems and Industrial Management (pp. 428–442). Springer, Berlin, Heidelberg.

[40] Gregory, R. and Mendelsohn, R. (1993). "Perceived risk, dread, and benefits." *Risk Analysis, 13*(3), pp. 259–264.

[41] Jiang, L., Chen, H. and Deng, F., 2010, May. A security evaluation method based on STRIDE model for web service. In 2010 2nd International Workshop on Intelligent Systems and Applications (pp. 1–5). IEEE.

[42] Peltier, T.R., 2000. *Facilitated risk analysis process (FRAP).* Auerbach Publication, CRC Press LLC.

[43] Ali, B. and Awad, A.I., (2018). "Cyber and physical security vulnerability assessment for IoT-based smart homes." *Sensors, 18*(3), p. 817.

[44] Keating, C.G., 2014. Validating the octave allegro information systems risk assessment methodology: a case study. 28

[45] Huang, G., Li, X., and He, J., 2006, May. Dynamic minimal spanning tree routing protocol for large wireless sensor networks. In *2006 1ST IEEE Conference on Industrial Electronics and Applications* (pp. 1–5). IEEE.

[46] Alghamdi, B.S., Elnamaky, M., Arafah, M.A., Alsabaan, M., and Bakry, S.H., (2019). "A Context Establishment Framework for Cloud Computing Information Security Risk Management Based on the STOPE View." *IJ Network Security, 21*(1), pp.166–176.

[47] Chou, D.C., (2013). "Risk identification in Green IT practice." *Computer Standards & Interfaces, 35*(2), pp. 231–237.

[48] Aven, T., 2015. *Risk analysis.* John Wiley & Sons.

[49] Masuoka, Y., Naono, K., and Kameyama, S., Hitachi Ltd, 2004. *Network monitoring method for information system, operational risk evaluation method, service business performing method, and insurance business managing method.* U.S. Patent Application 10/629,920.

[50] Sheehan, B., Murphy, F., Ryan, C., Mullins, M., and Liu, H.Y., (2017). "Semi-autonomous vehicle motor insurance: A Bayesian Network risk transfer approach." *Transportation Research Part C: Emerging Technologies, 82*, pp.124–137.

[51] Bhavsar, P., Das, P., Paugh, M., Dey, K. and Chowdhury, M., (2017). "Risk analysis of autonomous vehicles in mixed traffic streams." *Transportation Research Record, 2625*(1), pp. 51–61.

[52] Ullman, G. L., Finley, M. D., Bryden, J. E., Srinivasan. R., and Council, F. M., (2008) NCHRP Report 627: Traffic Safety Evaluation of Nighttime and Daytime Work Zones. Washington, DC: Transportation Research Board.

[53] Dezfuli, H., A. Benjamin, C. Everett, G. Maggio, M. Stamatelatos, and R. Youngblood. NASA Risk Management Handbook. Publication NASA/SP-2011-3422. NASA, 2011.

[54] Roy, A., Kim, D. S., and Trivedi, K. S., (2012) "Attack countermeasure trees (act):towards unifying the constructs of attack and defense trees." *Security and Communication Networks, 5*(8), pp. 929–943.

[55] O. Henniger, L. Apvrille, A. Fuchs, Y. Roudier, A. Ruddle, and B. Weyl,"Security requirements for automotive on-board networks," in 9th International Conference on Intelligent Transport Systems Telecommunications,(ITST), 2009, pp. 641–646.

[56] Petit, J. and Shladover, S. E., (April 2015) "Potential cyberattacks on automated vehicles." *IEEE Transactions on Intelligent Transportation Systems, 16*(2), pp. 546–556.

[57] Behfarnia, A. and Eslami, A., 2019, September. Local voting games for misbehavior detection in VANETs in presence of uncertainty. In *2019 57th Annual Allerton Conference on Communication, Control, and Computing (Allerton)* (pp. 480–486). IEEE.

[58] A. Boudguiga, A. Boulanger, P. Chiron, W. Klaudel, H. Labiod, and J.-C. Seguy, "Race: Risk analysis for cooperative engines," in 7th International Conference on New Technologies, Mobility and Security(NTMS). Paris, France: IEEE, July 2015.

[59] Society of Automotive Engineers (SAE), SAE-J3016: Taxonomy and Definitions for terms Related to Driving Automation Systems for On-Road Motor Vehicles, Sep 2016.

[60] G. Sabaliauskaite, J. Cui, L. S. Liew, and F. Zhou, "Integrated safety and cybersecurity risk analysis of cooperative intelligent transport systems," in SCIS-ISIS, Toyama, Japan, pp. 723–728, Dec 2018.

[61] Jeon, S.H., Cho, J.H., Jung, Y., Park, S. and Han, T.M., 2011, February. Automotive hardware development according to ISO 26262. In *13th International Conference on Advanced Communication Technology (ICACT2011)* (pp. 588–592). IEEE.

[62] The traditional ISO 26262 can be applied along with Hazard and Risk Analysis (HARA), System Theoretic Process and Analysis (STPA) and Automotive Safety Integrity Level (ASIL) for every components of AV separately, as for example for engine systems.

[63] Shastry, A.K., 2018. *Functional safety assessment in autonomous vehicles* (Doctoral dissertation, Virginia Tech).

[64] A. Nanda, D. Puthal, J. J. P. C. Rodrigues and S. A. Kozlov, (August 2019) "Internet of Autonomous Vehicles Communications Security: Overview, Issues, and Directions." *IEEE Wireless Communications, 26*(4), pp. 60–65. doi: 10.1109/MWC.2019.1800503.

[65] Fleetwood, J., (2017). "Public health, ethics, and autonomous vehicles," *American journal of public health, 107*(4), pp. 532–537.

[66] Duncan Langford (1997) "Ethical issues in network system design." *Australasian Journal of Information Systems,* 4. doi: 10.3127/ajis.v4i2.367.

[67] Kopelias, P., Demiridi, E., Vogiatzis, K., Skabardonis, A. and Zafiropoulou, V., (2020). "Connected & autonomous vehicles–Environmental impacts– A review." *Science of the Total Environment,* 712, p.135237.

[68] Takeyoshi Imai (2019) "Legal regulation of autonomous driving technology: Current conditions and issues in Japan." *IATSS Research, 43*(4), pp. 263–267. doi: 10.1016/j.iatssr.2019.11.009.

[69] Maurer, M. *et al.* (eds) (2016) *Autonomous driving: technical, legal and social aspects.* Berlin: Springer OPen. doi: 10.1007/978-3-662-48847-8.

[70] Grama, J. L. (2011) *Legal issues in information security.* Sudbury, Mass.: Jones & Bartlett Learning (Jones & Bartlett Learning information systems security & assurance series). Available at: INSERT-MISSING-URL (Accessed: December 6, 2020).

[71] Liu, N., Nikitas, A. and Parkinson, S. (2020). "Exploring expert perceptions about the cyber security and privacy of Connected and Autonomous Vehicles: A thematic analysis approach." *Transportation Research Part F: Traffic Psychology and Behaviour,* 75, pp. 66–86.

Chapter 7

Statistical in-depth security analysis for vehicle to everything communication over 5G NETWORK

Rejwan Bin Sulaiman
University of Bedfordshire, Luton, United Kingdom

Ranjana Lakshmi Patel
Capgemini, Paris, France

CONTENTS

DOI: 10.1201/9781315110905-7

7.1 INTRODUCTION

Researchers and companies have placed a lot of attention on the vehicular network and the impact of 5G in V2X communication. The ETSI is responsible for coming up with a standard V2X to shape up an ITS. The standards include a protocol stack, security mechanisms, security requirements, and architectural stability. (European Telecommunications Standards Institute, 2010).

For V2X communications, the 3rd Generation Partnership Project (3GPP) studied Long Term Evolution (LTE) to confirm it is possible to use it (European Telecommunications Standards Institute, 2004, 2009). Unlike LTE, "LTE Advanced" can fulfill IMT-Advanced requirements for the 4G. Again, to develop the standards, the ETSI has been responsible for the augmentation of LTE for V2X communications. (European Telecommunications Standards Institute, 2017).

On the other hand, the ETSI ITS standards have been adapted compared to the LTE V2X communication. Efficiency, speed, and reliability are facilitated in 5G NR technology. In 2015, improvements to 5G began. The standards were then re-organized to support the traditional mobile networks through multiple endnotes, including supportive vehicles and Internet of Things (IoT) (5G Automotive Association, 2016). High-speed and broad bandwidth can be achieved with the assistance of 5G NR technology.

That being said, researchers have not investigated, or put time and energy into, one crucial notion. Researchers have not paid much attention to finding out how 5G impacts the security of V2X communications. Naturally, with the introduction of new technology comes different types of security hurdles. Nevertheless, 5G can bring about new possibilities to protect and secure vehicles and provide security mechanisms in the ITS model.

This research can be viewed as a complement to the statistical in-depth security analysis for the V2X communication over the 5G system of the European Union (EU) 5G Infrastructure. It can also be related to the different types of V2X projects at the Chalmers and the FFI/Vinnova projects. Volvo and the Corporation of Volvo Car concentrate on augmenting the privacy and the security for future vehicles, which will be run and operated by future generations. Hence, this fundamental research will amalgamate researchers' viewpoints on in-depth security analysis on 5G for V2X communication and ITS stack security.

7.1.1 Motivation

At present, 802.11p is being used in all link-layer protocols concerning the V2X communication system. However, this presents many hurdles, such as it lacks security to begin with. Given 5G is advancing accordingly with the times, thousands of companies and researchers are looking at using V2X communication with assistance from 5G technology. Some features do not resemble the 802.11p in 5G NR, and provide the component with new possibilities. In the meantime, however, vulnerabilities are evident from the other side. To be more specific, the executed security mechanisms belong to the higher end, which holds the possibility of being exchanged with the functionality of 5G. Hence, the implemented mechanism might become useless or redundant at times. Nevertheless, many features will provide upgraded security and protection to the people placed in this research.

7.1.2 Research questions

Five research questions have been formulated to explain the research in an organized manner:

- RQ1: What types of security mechanisms and protocols have been considered in ETSI V2X communications until now?
- RQ2: What type of security requirements are needed for several use cases of ETSI ITS?
- RQ3: What features can be utilized at the physical layer to develop 5G security measures? Is it possible to repudiate any security mechanism from higher-layer protocols?
- RQ4: What are the vulnerabilities that 5G NR can create for V2X communication?
- RQ5: Will it be possible to replace 802.11p with 5G NR? What are the obstacles if it is not possible to replace them?

7.1.3 Aim of the research

This research investigates the statistical in-depth security analysis for the V2X communication over the 5G network and the security effect of V2X communications. To improve security and vulnerabilities, we have analyzed and examined different views on the research topic. Additionally, essential reductions are explained, which are needed to define the protocol stack accordingly. This research aims to find out whether 802.11p can be replaced using the 5G. Neither associated problems, nor the solution of the problems, have been taken into account.

Ultimately, the research does not have any intent to provide the best optimization system for all security mechanisms. However, the goal is to collect the information so that all possible optimization can be localized.

7.1.4 Contribution

This research focuses on a statistical in-depth security analysis for V2X communication over the 5G network and how 5G can affect communication security. In this research, it has been investigated that it is not impossible to make a transit from the 802.11p to 5G NR in terms of security aspects. However, it needs some alteration in the protocol stack. That is why different types of improvements have been proposed, so that the security mechanisms will get stronger and better. Advancing technology will eradicate the present certificate mechanism, which is needed for authentication. This will be done only because the prospects and aspects of 5G will provide the communication with so many features that authentication will not be needed. Towards the end of the chapter, we discuss future recommendations and the positive and negative sides of V2X communication.

7.1.5 Scope

This research has limitations given it has been restricted by different resources, such as time and data availability. Scopes of doing the study are as follows:

- This research can be considered a general study that will not possess a detailed analysis of the execution of several types of protocols. In this research, protocol stack and associated topics are discussed to provide an overall view.
- The abstractions have been originated from the execution of 5G NR and 802.11p protocols. In terms of security perspective, this research only provides pertinent information to investigate the differences between them.
- Given time has been an issue, it was impossible to include a new protocol stack, such as integrating 5G into the ITS stack.

7.2 LITERATURE REVIEW

The technological transformation occurs in the automated vertical market, and is intended to create more concentrated vehicles with fully automated capabilities. It is possible to do so with the assistance of 5G. This can be done based on the connection between the autonomous vehicles (AVs) incorporated with each other and by considering the Vulnerable Road Users (VRUs), V2X in 5G network systems. V2X is capable of formulating the perception of the environment. Thus, it can make the right decisions by exchanging views among the vehicles close to the environment. The automated vehicles can ensure a safe transportation system, which can reduce the fatality rate minimally. Also, traffic congestion will be reduced, as will the impact on the environment. However, the challenges persist. To find the answer to the research questions, 5G, 5G NR, ETSI ITS, and the Institute of Electrical and Electronics Engineers (IEEE) Wireless Access in Vehicular Environments (WAVE) standards, and cellular V2X are discussed.

7.2.1 The vehicular network standards

7.2.1.1 ETSI ITS

Different types of standards have been devices with regard to ITS communication (ITSC). The standards have inclusion regarding the protocol stack(Gianotti et al., 2016); the communication architecture (Visala, 2014); security requirements as well as the messages (Visala, 2014), and the architecture and services. The end nodes in those standards are known as the *ITS station*. The vulnerability and the execution assessment is be done in the sections that follow.

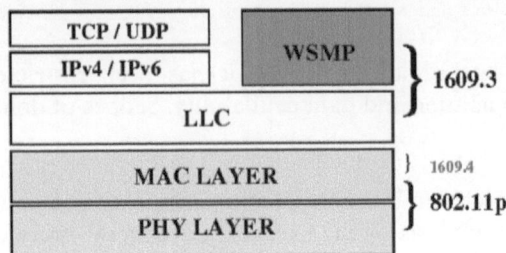

Figure 7.1 WAVE standard protocol stack.

7.2.1.2 IEEE WAVE standard

WAVE has been introduced in the United States by the IEEE based on V2X communication (Ahmed, Ariffin, & Fisal, 2013). WAVE has been undertaken to demonstrate several standards that are needed for V2X communication. With respect to security and architecture, WAVE is quite close to the European standard.

Figure 7.1 is a WAVE stack protocol. The difference can easily be discovered in the upper layers given the WAVE Short Message Protocol (WSMP) is utilized instead of the GeoNetworking protocol or Blockchain Transmission Protocol (BTP) in ETSI ITS standards. According to Ahmed et al. (2013), this protocol is used for an instant inter-exchange of packets that reduces communication overhead time. WSMP and User Data Protocol (UDP) services are not clearly defined in the WAVE services. Different types of cryptographic mechanisms, such as the public/symmetric keys for signing/encryption, verify the certificates given by the authority's infrastructure. The certificates can be one of two types, such as *explicit*, which means that the public key is explicit, and *implicit*, which includes a reconstructed value for the public key.

7.2.2 5G for V2X

Much research has been done with a focus on optimizing V2X communication. According to Aurora et al. (2010), the authors present a new proposal for non-orthogonal time-frequency schemes for allocating high reliability and low latency. A testbed creation was performed for low latency and high-reliability 5G-V2X communication systems (Ali et al., 2017). Another proposal from Wooguil Pak et al. (2017) states that a faster and smoother pack of classification can enhance the performance. According to Zhou and Kellerer, (2017), Virtual Cells (VCs) can optimize the efficiency of power, reliability, and capacity. However, Luoto et al. (2017) finds that optimum performance can be discovered between the vehicles' LTE communication, which possesses a comparatively sophisticated connection server with the Roadside Units (RSU). As described in Boban et al. (2016),

Global Navigation Satellite System (GNSS)-based synchronization has been one of the most important network coverage scenarios. The multi-antenna can provide good coverage and necessary developments related to Ultra Reliability and Low Latency Communication (URLLC). According to the ETSI ITS definition, in testing, BISSender, BISReceiver, and different layer utility messages were used by the software programs (European Telecommunications Standards Institute, 2011). The whole pack is then delivered to the Radio BBU as a means of UDP messages, sent by the 5G communication device to device. The rates of the messages went up to 200 Hz, which is usually kept between 1 Hz–10 Hz.

7.2.3 C-V2X vs 802.11p

The research questions contained the possibility of whether it is possible to convert from 802.11p to 5G NR. It is important to find the differences between them so that the comparison can be made easily. Given 5G NR is still in development, the comparison will be between the 802.11p and Cellular V2X (C-V2X). Table 7.1 depicts this comparison:

However, in the 802.11 protocol, this has been devised solely for the ad-hoc mode. Different vehicles can connect and communicate directly. Before setting up a connection, adding a Basic Service Set (BSS) is recommended. Though it might not possess any security in 802.11p, the reliability is on a higher security mechanisms level. Additionally, according to Filippi et al. (2017), the 802.11p is not sophisticated or well-structured, like C-V2X.

Table 7.1 Comparison

LTE-V2X	802.11p
Download speed is up to 100 Mb/ps Upload speed is up to 50 Mb/ps	The range is from 1.0 to 63.5 Mb/ps
Bandwidth (MHz) in term of channels: 1.4, 3, 5, 10, 15, 20	Channels of 10 MHz bandwidth in 5.9 GHz
OFDNA and Single Carrier–Frequency-Division Multiple Access (SC-FDMA) are used to download and upload respectfully	Carrier Sense Multiple Access with Collision Avoidance (CSMA/CA) methods are used for access
Quadrature Phase Shift Keying (QPSK), 16 Quadrature Amplitude Modulation (16QAM), 64QAM are used for modulation	Binary Phase Shift Keying (BPSK), QPSK, 16QAM, 64QAM
CRC, Hybrid Automatic Repeat Request (HARQ) used for error correction	CRC, Forward Error Correction (FEC)
Half duplex, Frequency Division Duplex (FDD), and Time Division Duplex (TDD)	Only half duplex

7.2.4 5G NR (New Radio) frequency

5G has the potential to develop a new technology based on radio usage. This is called NR. NR has been devised so that communication can take place with different types of systems. At the end of 2017, NR was released primarily on 3GPP.org. This provides credible and reliable communication by consuming broader bandwidth. Some modern technologies have been proposed by Satrya et al. (2017) for the 5G physical layers, such as Visible Light Communication (VLC), Millimeter Wave (mmWave), Non-Orthogonal Multiple Access (NOMA), massive multiple-input-multiple-output (MIMO), and cognitive network. Shimaa et al. (2020) introduced recent usage of 5G-V2X's normal development with particular security aspects. Researchers discussed the potential takeover of data, or packets, from one vehicle to another transmitting a secret key. Upon sharing the secret key, both the vehicles can agree on a common light-weight algorithm to secure the packet that is transferred among each other. These algorithms have been selected to secure device to device (D2D) communication.

7.2.5 mmWave

The LTE radio possesses an optimum frequency of up to 2.6 GHz. However, there is a need for greater data-transmission rates. To meet this need, the mmWave can provide a huge spectrum of frequency from 30 GHz to 300 GHz by utilizing a minimal wavelength of 1mm to10 mm, as described by Boban et al. (2016). Boban et al. (2016) also suggests that the antenna size has to be decreased, as does the size of the wavelength. Further, short wavelengths may create hurdles given they do not work in harsh weather. To solve the matter, the 5G NR can include multi-node beamforming technology.

7.2.6 MIMO and beamforming in 5G

Using beamforming technology can add reliability to mmWave technology to a great extent. Given antenna sizes are small, it is possible to have several separate antennas in the base station. The MIMO antenna enables concentrated beams to aim at the users who are individual in nature. Using beam tracking and beam training, a beam can efficiently provide users with an uninterrupted communication setup through identical links, as published at an IEEE conference (Figure 7.2).

The technologies mentioned previously have been investigated with regard to V2X communication. Va et al. (2016) posits that high-quality mobile devices are needed for faster beam training. Previously, many beam pairs

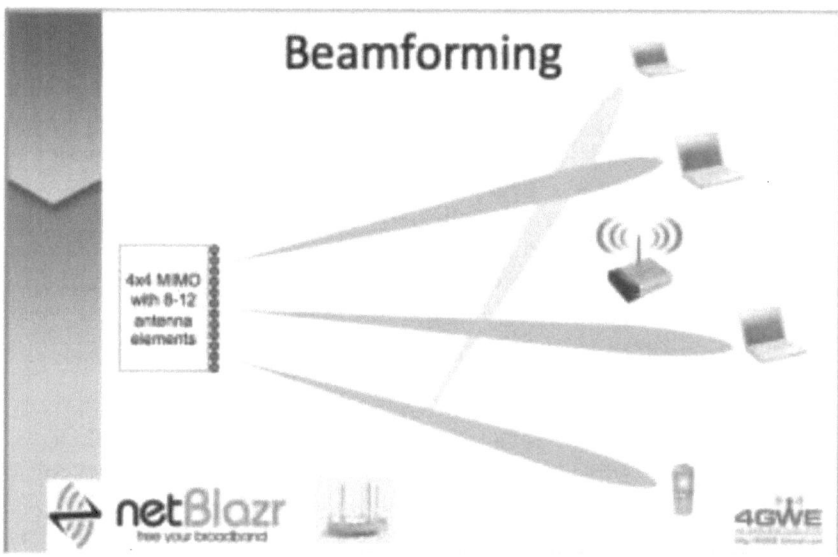

Figure 7.2 A visualization of how beams can be formed using massive MIMO antenna arrays.

had been used instead of a single beam. Larsson et al. (2017) investigates the same phenomenon by executing a proficient tracking of beam pairs in volatile situations where the frequency switches remained between all of the transmission points. Researchers conducted an experiment on a Bavarian Motor Works (BMW) racetrack, which made them more enthusiastic. They placed emphasis on the control of Doppler shifts introduced by high speed. Tateishi et al. (2017) proposed a solution that states that Channel State Information (CSI) can be selected to minimize the overhead of signaling. Zhao et al. (2017) came up with an idea for securing the communication system. MIMO signaling, as well as the duplex operation, is supported by the two-way communication system. A paper by Al-Momani et al. (2016) proposed using beamforming so that the re-authentication of the physical layers can be augmented. A channel called "signature" is used for the authentication, which is an amalgamation of different Doppler shift variations.

7.2.7 Visible light communication (VLC)

VLC is attracting researchers all around the world. VLC is regarded as radio transmission, which is not a significant portion of the 5G NR. This technology is used in the Reasonable Optical Near Joint Access (RONJA) project as part of its early execution in 2001, according to ronja.twibright. com (2017), and has provided better executions of 5G developments.

VLC is responsible for transferring data rapidly by flickering light-emitting diode (LED) light, which the naked eye cannot comprehend. If the light is switched on, it will portray 1, and if it switched off, it will show 0, respectively per Karthik et al. (2017). With regard to backhaul communication, the VLC system can add value between the national core and cell towers. The benefit of the VLC system lies in its distance outdoors. However, there are definitely some challenges executing the VLC system in the long run. One of VLC technology's benefits is it is cost effective and more environmentally friendly when compared to mmWave. VLC technology challenges, or drawbacks, is it is more expensive due to its rigorous purpose, and, therefore, operating expenses is enormous.

In contrast, VLC does not require any expensive devices within itself. However, drawbacks can include the low Signal-to-Nose Ratio (SNR) so that the light can widely disperse and remain sensitive to surroundings. Haas et al. (2020) introduces hybrid Light Fidelity/Wireless Fidelity (LiFi/WiFi) on the train. They call it a "grey system." The takeover of the data transmission wis decided from the central server. The signal and interference ratio is the key decision element.

7.2.8 Non-orthogonal multiple access (NOMA)

NOMA is another method used in the 5G NR, and is discussed in Kizilirmak (2016) (Figure 7.3). Time Division Multiple Access (TDMA), Frequency Division Multiple Access (FDMA), and Code Division Multiple Access (CDMA) are methods that can add value to NOMA. However, they fall under Orthogonal Multiple Access (OMA) methods and cannot provide a 5G NR cooperation requirement.

In NOMA, several users can transmit simultaneously using an identical frequency. The signals can be transmitted in the form of a wave by separating the receivers' power level. The separation is regarded as the Successive Interference Cancellation (SIC). Since the separation needs to be iterative in nature, it is known as the *successive*. The topmost strong signal is taken for extraction and subtraction until the desired signal is revealed. Spectral Efficiency (SE) and Energy Efficiency (EE) can provide a high data rate, which can only happen provided that the cancellation

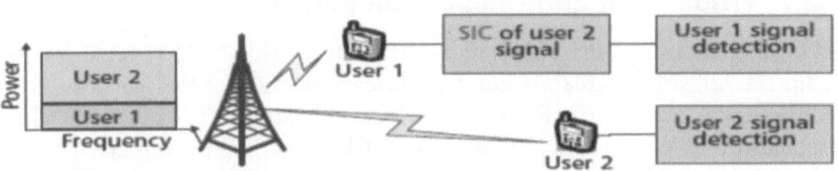

Figure 7.3 NOMA method.

is perfect after the assumption of SIC. However, in real life, it is virtually impossible that the perfect cancellation can take place given some sort of interference would occur. This is why the data rate is lower, due to remaining interference. In fading channels, errors within the cancellation are expected. One of the problems of using NOMA is the high computational power requirement for the SIC algorithm to run many users. For the vehicles, it will be a hurdle to optimize the allocation of power. To control the challenges, it is imperative to accumulate MIMO with NOMA to minimize the errors. The combination will increase the reliability. Privacy is one of the NOMA concerns investigated by Satrya and Shin (2017). They addressed that iterative separation by SIC can enable a harmful user to eavesdrop. The proposal's solution includes the importance of two keys, which should be used to eliminate the problem. It means that one key should be used for the hashing, whereas the other key should be used for authentication. The first key is responsible for making sure that the legal user is performing. In contrast, the second key consists of the embodiment of the medium access control (MAC) and International Mobile Equipment Identity (IMEI) of the users.

7.2.9 Cognitive radio (CR) networks

Soliman et al. (2017) states that cognitive radio (CR) can play a crucial role in augmenting the radio spectrum's performance (see Figure 7.4).

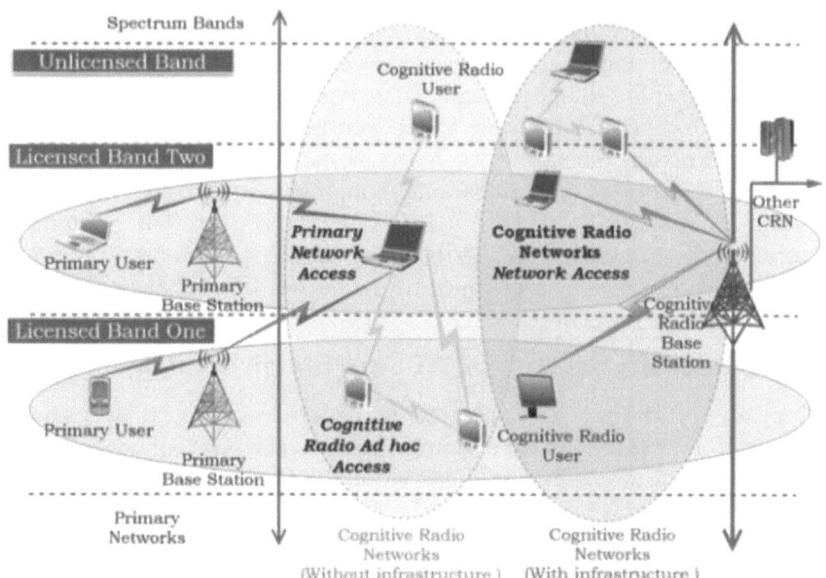

Figure 7.4 CR network mechanism.

It is responsible for bridging licensed and unlicensed users so they can coexist by utilizing various spectrums. Licensed users are a higher priority and also are regarded as *primary* users. On the other hand, unlicensed users are regarded as *secondary* users. Secondary users can only transmit when the spectrum possesses a hole left off by the primary users. Software Defined Radio (SDR) can help achieve the spectrum-sensing radio given it is way more convenient for the user (Soliman et al., 2017). This particular technology possesses improvements in capacity as well as throughput rate. However, the technology's complex operation make it vulnerable to attacks (Soliman et al., 2017).

The high risk is on the primary user due to the primary and secondary users' coexistence. The chance of eavesdropping is not minimal. To address this problem, Xie et al. has come up with the idea to add noise in the signal through the Superimposed Coding (SC) to create an interference in the original signal; hence the primary user's concern for eavesdropping is solved. With regard to SDR, Grigorios Kakkavas et al. (2020) proposes an implementation framework and demonstrates the Markov random field (MRF) framework's feasibility. This provides a solution to the transmission protocols in the CR Networks (CRN).

7.2.10 The security aspect of 5G in V2X

Bian et al. (2017) discusses security aspects of the V2X communication system. They address other issues as well, such as a platoon disruption attack that they identified based on old packets and data distortion attacks by providing false pictures or videos. On another note, it was almost impossible to find the jamming attacks. These issues can be addressed by introducing non-cryptographic security mechanisms. An example can be taken from a specific platoon car responsible for collecting data from different vehicles to understand any deviations, such as statistical interference. To get rid of jamming, it is vital to set up a channel-hopping process so that the communicating parties can hop on through a sequential process that is completely exotic to the attacker. One of the most challenging non-cryptographic mechanisms is the delay that concerns the parties (Eiza et al., 2016).

7.2.11 Security aspects of 5G over the physical layer

Many investigations lead to research on the wireless physical layer security (PLS) in 5G. Researchers have identified two layers: physical layer authentication and PLS. According to Basem et al. (2019), PLS has been used in 5G technology in terms of confidentiality. The endpoint receiver channel quality remains good, even after receiving artificial noises. Studies have found that the cryptographic approach is slower than PLS technology. Hence, instead of a cryptographic approach, scientists are viewing PLS as a higher security solution.

According to Basem et al. (2019), 5G's latest products do not go hand and hand with the existing PLS technology. A solution has also been recommended to better fit the PLS technology with 5G features. In ultra-low latency, it is crucial to use PLS depending on where the eavesdropper does not get what he or she wants to get in the specified time.

That is why Farhang et al. (2015) proposed PLS that can protect the vulnerability in 5G NR. The access points are comparatively smaller, and that can actually play a part in disclosing the location. However, the location can be disclosed with pinpoint accuracy. The authors have proposed an interesting solution by formulating a mechanism to use the "noise" at the access point. The same thing has been found by Yu et al. (2016), who propose a solution through knowledge distribution.

Pan et al. (2015) has developed physical layer authentication based on channel response and unique. The channel's response can be used as a means of a fingerprint because of the uniqueness of the features. Xieet al. (2017) has also studied this in terms of authentication of the symmetric key in the distribution process.

7.3 METHODOLOGY

Research methodology is a vital component of research. In this section, we discuss theoretical and systemic analyses applied for research purposes.

Selecting and implementing a specific methodology is very important because it helps with data collection for literature and assembling facts. Irony and Rose (2005) used the method for research shown in Figure 7.5. Their flow chart demonstrates how researchers can go through data until they reach a particular result.

It makes the process simpler and helps observe new facts and draw conclusions. There are different types of research methodologies. And they differ depending on the research type and criteria. In this chapter, we are

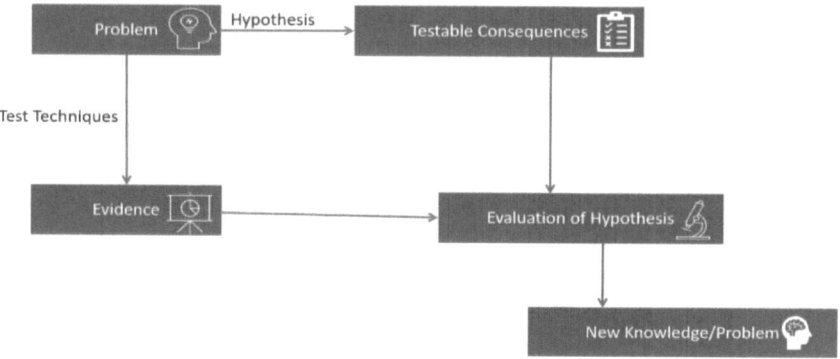

Figure 7.5 Flow chart proposed by Imy and Rose (2005).

mostly focusing on *descriptive research*, where we collect data from the different resources and other papers. In this section, we briefly describe how we collect data, how the research questions are resolved, or how we have try to solve each problem.

7.3.1 Research philosophy

There are several options to choose among with regard to a data collection system, based on the research topic or the research arena. Hence, it is a must for a researcher to follow a particular course of action to conduct the research. Although multiple choices of modes are given in the (Saunders et al., 2009) book, many have been made with concern for this study's needs.

It is necessary to have a discussion on the research Onion before getting into the details (see Figure 7.6).

The research onion portrays the research's step-by-step process, which entails the researcher's activities and the researcher's views. It is mandatory to go for an in-depth analysis. It is important to go through the layers of the onion thoroughly until the desired outcome is achieved.

Saunders (2009) introduces four philosophies on the research process, which let the researchers utilize one philosophy: incorporating two philosophies technically. The Pragmatic research philosophy depicts the research process in a multi-dimensional way and interprets the process of that research. The data include two important paradigms: *positivism* and *interpretivism* (Saunders et al., 2009). The positivism research paradigm relies on collecting data, whereas the Interpretivism paradigm focuses on the gathered data source. The interpretivism research paradigm can pave the way for analysis to achieve a possible solution (Saunders et al., 2009).

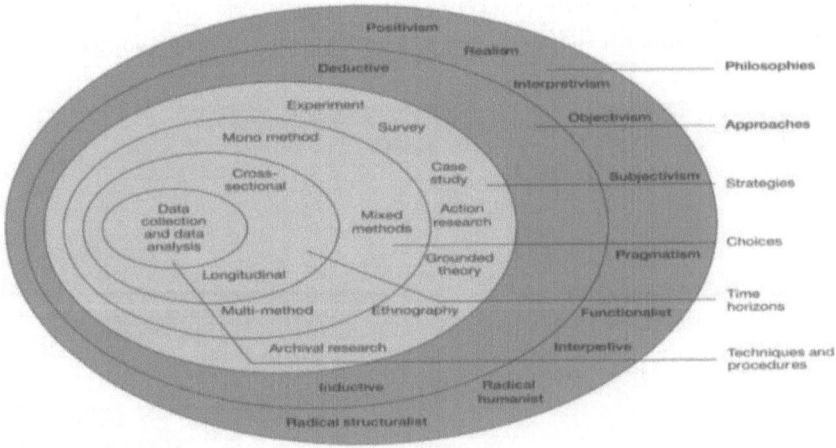

Figure 7.6 Research onion proposed by (Saunders et al., 2009).

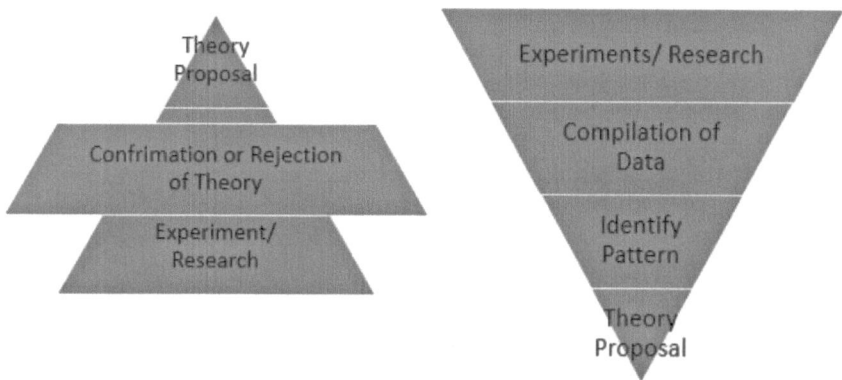

Figure 7.7 Research approaches (Creswell & Clark, 2007).

Two kinds of approaches are applied for reasoning that emphasize many parts of research (see Figure 7.7). One is the *inductive* approach, and the other is a *deductive* approach (Trochim, 2006). The first one is a process that takes a step from a very specific perspective to a general perspective. The second one, deduction, is a process that works with a general viewpoint and gradually moves toward a specific one. Deductive researchers usually begin their process using approach, such as a "'top-down" manner (Creswell & Clark, 2007).

Hence, the deductive approach can be summed up as a top-down approach, and it needs to possess a theory that is requires proof to be adopted. In contrast, the inductive approach is represented as a bottom-up approach (Saunders et al., 2009).

Alternatively, with the inductive research process, researchers tend to work from the bottom to the top while looking at respondents' opinions to create a wide array of themes and generate them according to the theory (Creswell & Clark, 2007). The approaches mentioned previously have similarities with their quantitative and qualitative methods, also known as deductive and inductive methods. But not every researcher supports the combination of inductive-deductive or quantitative-qualitative. Thus, it is prudent to treat them each differently.

Data collection and organization will be based on surveys, interviews, literature reviews, and questionnaires. Afterward, the organized data will be utilized to identify this research's aspects, enabling the researcher to deduce the research's main purpose based on the smart vehicular transportation system.

The researchers undertake numerous strategies that suit their research topics. The grounded theory, survey, ethnographic strategy, experiment research, action research, and combination of action and experiment research are some strategies to name a few. The research strategy mode will need to be justified to move forward with the research strategy (Bernard, 2011).

However, time can be of the essence with research, and some strategies require a lot of time to adequately address. This is why a survey can be designed to collect the primary data from the respondents, which is not very time-consuming, provided that the respondents are ample in number (Saunders et al., 2009).

This type of study requires collecting data from respondents who frequently use transportation to go from one place to another. Hence, it would not be wise to interview all people and, also, doing so would take a huge amount of time. Research strategies like action research, archival research, or grounded theory would not be suitable for V2V communication. At this time, there is not much information on the internet due given the concept is fairly new. Going for the survey would be the right fit to raise awareness among people regarding security concerns and collect the respondents' insights on their needs.

7.3.2 Data collection methods

In the primary data collection method, the difference can be understood by the data collection type. One type, quantitative, is mostly based on numerical data. In contrast, the other type, qualitative, represents the research's exploratory nature (McCusker & Gunaydin, 2015). *Quantitative* research concentrates on statistical data and is used by researchers in great density. It can provide exact information about what the researchers are actually looking for. On the other hand, *qualitative* research focuses on attitudes, and different correspondents and experiences to shed light on the research topic.

7.3.2.1 Interview

In an interview, data is collected by the researcher, who asks a respondent certain questions, and the respondent answers them accordingly. An interview is considered very important in an expert's consultancy because the expert can enrich the interview with valuable insights. The obtained data is of high quality given it comes from the respondent's experiences. In addition to this, the related cost of collecting data by arranging interviews is higher than other methods of data collection (Odom and Chinedum E, 2014).

We have conducted interviews with our fellow students and some cybersecurity experts in the United Kingdom for our research. The interviews have helped us come up with different research questions. Given the interview was between the people in the field, the questions were technical. We contacted some of the researchers around the world who are working on this topic. We sent them interview questions and asked them to answer them on their time.

One researchers from Sweden replied with in-depth answers over Skype, which helped us analyze our data.

7.3.2.2 Questionnaire

Another easy method of collecting data is a questionnaire. It represents the respondents' observations by asking them questions to express their thoughts and opinions regarding certain phenomena.

In our research, we have collected interviews from prominent researchers and associated personnel from different sources. This has provided important input in helping us come up with the 5G network's prospects concerning vehicular cellular network technology.

We include the questions we asked the researchers here.

The topic of interest consisted of, but was not limited to, the following:

- Access authentication for autonomous vehicular net
- Attack resistance
- Requirements for the ITS
- Security in ETSI v2X communication
- CIA triad model in the V2X network

7.3.2.3 Survey

Conducting a survey is a primary part of research. In our case, we used the case study as our research strategy. But we have also taken into account secondary survey data from surveys conducted by other researchers. This type of research type works as a medium to understand others' perceptions and analyze them to know more about a specific occurrence.

A survey technique's implementation and capacity cannot be restricted to a single area of focus because of that technique's characteristics. The survey is one of the most frequently used methods, and to conduct it requires a lot of care. Surveys can be used to understand the perspectives of the respondents. They can uphold the present scenario regarding the research topic. In the data analysis portion of our research, we used responses from the survey to conduct analysis.

7.4 ANALYSIS

In prior sections of chapter, we discussed the impact of 5G radio technologies in V2X communication, and how they can protect the existing system. In this section, we cover the data analysis given we already collected the data from the respondents. It is essential to mention that the data analysis is able to answer the research questions.

7.4.1 Analysis based on our research questions

In this section, we explain the extensive literature review used to determine the answers to the research questions posed at the beginning of the research.

The analysis is based on the CIA triad. That ensures the maintenance of the security, and also, that the goal of this research is achieved.

The questions are as follows:

- RQ1: What type of security mechanisms and protocols has been considered in ETSI V2X communications till now?
- RQ2: What type of security requirements are needed for several use cases of ETSI ITS?
- RQ3: What features can be utilized at the physical layer to develop the security measures of 5G? Is it possible to repudiate any security mechanism from the higher layer protocols?
- RQ4: What are the vulnerabilities that 5G NR can cause to V2X communication?
- RQ5: Will it be possible to replace 802.11p with 5G NR? What are the stymies if it is not possible to replace them?

Security stands for the protection of vital information to control the accessibility of unauthorized personnel. It actually is a combination of Confidentiality, Integrity, and Availability.

The research questions are addressed, and answered accordingly, with literature reviews and analysis of the survey responses.

7.4.1.1 RQ1: Security mechanisms and protocols in ETSI V2X communications

According to section 7.2.1, this particular question has already been answered and explained in full. The findings are summarized here. The security mechanism used in the ETSI ITS will be described in brief in the following text. The packet of Secured Message is a combination of different types of payloads that are DENM or CAM in nature. These are the requirements that the security layer needs for the various types of trailers and headers. The Secured Message header carries a unique version protocol as well as the security profile. The security profile is used to portray the messages' format, such as payload, compulsory header, and trailer fields. Further, it is also used in terms of encoding.

In the case of **confidentiality**, CAMS and DENMs have stated that no messages should be encrypted as per the security profile for the CAMs and DENMs (European Telecommunications Standards Institute, 2013). As the research moves towards the following sections, data confidentiality can merely be regarded as a problem in applications for the basic sets.

At the time of **authentication** of any user through a certificate authority, a public key is being used for exchanging the credentials cryptographically (European Telecommunications Standards Institute, 2010).

On the other hand, authentication and **integrity** in the ETSI ITS model experience the cryptographic signatures so that the sender's authentication can be done. The integrity of the messages can be guaranteed (European Telecommunications Standards Institute, 2010). The message is verified and signed using asymmetric cryptography or symmetric cryptography.

Privacy is guaranteed for the sender given the user cannot be tracked down. Hence, the pseudonyms must possess updates every now and then (European Telecommunications Standards Institute, 2012). Besides, it is tough and close to impossible for the attacker to establish a connection or a link of the pseudonym to the canonical identity of the users (European Telecommunications Standards Institute, 2010).

No guarantees of availability have been found, which can explain the ETSI ITS security documents. According to European Telecommunications Standards Institute (2010), availability has already been taken into account along with the solutions. It includes messages like "Include the origin address in every V2V message," "Limitation of all message traffic to the V2I/I2V and the places where it is suitable," and "Include a sequential number in the new message." One of the certificate structure requirements is to provide the validation and distribution of the asymmetric keys so that the encryption and the signatures can be done. Granting the certificates is a responsibility of the Certificate Authorities (CAs) who does the work by maintaining a hierarchy.

The "ultimate root of trust" appears at the top of everything, and it is called the Root CA. In contrast, the down livings are called the Authorization CA and the Enrollment CA. Enrolment CA is responsible for the inclusion of the initial authentication. It is also responsible for providing pseudonyms to users. Afterward, the Authorization CA provides the users with permissions relating to their needs to perform operations accordingly.

7.4.1.2 RQ2: Security requirements are needed for several use cases of ETSI ITS

7.4.1.2.1   Availability

This application's class has very rigorous requirements on availability given the safety concern is regarded as one of the highest priorities to be delivered.

7.4.1.2.2 Confidentiality

Confidentiality is not very important because warnings are delivered to those who might be vulnerable to upcoming dangers.

7.4.1.2.3 Integrity

In terms of safety measures, it is vital to acknowledge integrity. A fabricated message has the potential to divert the operation in a wrong way. For example, if a vehicle sends a warning on the "slow vehicle," it should mean only that. However, suppose it is modified and means something that is not very urgent, like a Point of Interest notification. In that case, this might lead to havoc on the road.

7.4.1.2.4 Authentication

Authentication can vary among numerous cases. In the case of warning messages, it is only a requirement from the receiving transports to identify the location. With concern for the receiving end, no actual need for the identity persists because of the authenticity of the users' location and the delivery of the messages. On the contrary, it might be used for audit purposes so that an identification can be made for misbehaving users. However, it is possible to falsify the identification. The attacker can send a fabricated warning for an accident. The receiver will be able to notice that the sender possesses another location through the location authentication. That is why it is not important for full authentication. Nevertheless, special vehicles must have authentication, such as emergency vehicles using the emergency vehicle warning system

7.4.1.2.5 Privacy

Privacy is very important, and the requirements can vary at times. Privacy provides protection and security for the slow vehicle indications, emergency vehicle warnings, co-operative glare reductions, and motorcycle approaching indications. These are the most used messages delivered during the trip. In the case of a collision warning, a privacy risk may arise. Privacy might seem to be problematic for the lane change warning and the overtaking vehicle warning. The user frequently overtakes and changes lane. These cases comprise road hazards warnings, and they have no potential threat of tracking users.

7.4.1.3 RQ3: Security measures of 5G in New Radio (NR) frequency

Except for the emergency vehicle warning, authentication is required for the receivers to locate the sender's location. It is also important to have the authorities' identity to remain in the business of tracking misbehaving users.

7.4.1.3.1 Physical layer authentication

In the physical layer, location authentication can occur by utilizing the physical layer authentication technology. Further, beamforming technology can augment technology and has the potential to make life difficult for the attacker to anticipate the location-dependent channel signature. It can also provide better performance for the collision risk warning. But this technology is not suitable for decentralized floating car data given the sender is not required to stay at a particular place.

7.4.1.3.2 Availability through NOMA and cognitive radio

In general, for any road safety cases, availability is a must. A special risk occurs in the decentralized floating car with regard to availability. The messages are repeated numerous times from a particular time to another car, and an attacker has the potential to multiply the messages and turn the messages into *waves*. As a result, a message flood can occur, which eventually jams the network. Consequently, users will lose access to the network. This is why it is imperative to install mechanisms to work as a shield to encounter the attacks. CR networks, as well as NOMA scheduling, can provide availability upon being utilized. Besides, as per sections 7.2.8 and 7.2.9, these technologies can efficiently utilize the network's availability.

7.4.1.4 RQ4: vulnerabilities that 5G NR can cause to V2X communication

7.4.1.4.1 Vulnerabilities

Some security requirements can be solved by the new security solutions discussed in the prior sections. However, the 5G NR is prone to vulnerable security attacks, which means the OEMs should be on their feet to tackle the vulnerabilities efficiently. The following sections provide solutions that have been identified based on the responses of those interviewed.

7.4.1.4.2 Privacy

New technologies appropriate to the 5G NR are discussed in this chapter. It has been noted that privacy can be a concern. The user's location can be found by introducing small cells that force mmWave to work for the attacker. According to Section 7. 2.11, this can be prevented by doing a randomization of the base. In Section 7.2.6, it has been discussed that privacy can be an issue in using beamforming technology given this requires tracking the user's location. NOMA can originate privacy concerns given the receiver is responsible for extracting the strongest signals with different destinations, as discussed in Section 7.2.8.

7.4.1.4.3 Availability

Availability problems can be found in the CR since the operation is complex, according to section 7.2.9. Proposed methods in that section can ease the complications of the CR.

7.4.1.4.4 Confidentiality

Confidentiality can be a concern in the CR. Therefore, many authors have studied, and come up, with numerous solutions on confidentiality issues in the CR, which are discussed avidly in section 7.2.9. One of the solutions was to introduce noise to the eavesdroppers. In the meantime, the channel between the receiver and the sender remains unharmed. This method is identical to the PLS.

7.4.1.5 RQ5: Possible exchange of 802.11p with 5G NR

In the beginning, our research question posed if the replacement of 802.11p with 5G NR might be possible. To answer that, it is necessary to acknowledge that it is possible to work simultaneously. However, the NR has not been defined, and it is possible to improve it according to the needs. Because of the modularity network stack, the change of scheduling method (NOMA) or frequency transmission should not be able to put any impact on higher layer protocols in this regard. According to subsection 7.2.3, the 5G NR is regarded as the developing technology. Hence, it is not be wise to compare this with the IEEE 802.11p protocols, which lay in the ETSI ITS model. However, the 802.11p protocol can be compared to C-V2X. In another subsection of 7.2.3, the table shows the technologies features mentioned previously. The C-V2X and the 802.11p possess almost identical features. The differences are noted in Table 7.2.

According to the table, the stymies of the C-V2X can pose a serious threat to the benefits. Having said that, the problems have already been solved the problems in 5G V2X. Besides, the Doppler effect has also been considered. It has also been solved given this effect can become problematic when the vehicles move fast. This actually represents the connection of both the LTE and 5G NR base station (Qualcomm. Designing 5G NR, 2018)

7.4.1.5.1 Redundant cryptographic mechanisms

We are portraying every aspect of the security concern. The security concerns of the use cases are integrity, availability, and authentication. Additionally, it is also understood from the discussion that the physical layer technologies

Table 7.2 Benefits and drawbacks of using C-V2X instead of 802.11p (Quadcomm, 2018)

Benefits	Drawbacks
More productive when the load goes up	Synchronization is not easy
Less collision because of the scheduling and load balancing infrastructure	Bad in collision handling
No hidden node problem	Near-far problem
Flexible in getting resource	The vulnerability of the dropper effect
More extensive coverage	Less coverage
	Frequency errors

can assist in giving necessary solutions. The receiver needs the authentication of the location. This is exactly where the cryptographic signatures can prove to be redundant in the use cases. However, this problem can be taken into action by channeling the signatures. Further, encryption is not needed for the selected use cases. Cryptographic authentication can be amalgamated with the physical layer authentication to achieve full authentication in identity and location [72], [15]. The certificate infrastructure and the cryptographic keys are needed in particular cases. It is done when the authority wants to track the users who have been behaving differently than they should be. If Identity-based Cryptography (IBC) can be embraced for the vehicular system, the certificate infrastructure can head towards being redundant. Earlier sections discuss the IBC in relation to the IoT. However, it is necessary to have adjustments for the V2X. Needless to say, this technology has the potential to become one of the most important technologies in the future that can ensure the optimization of the cryptographic mechanisms. Thus, it is an essential sector to devote resources to.

7.4.1.5.2 Other 5G security technologies

As previous sections advised, integrity and authentication can have a solution to a small extent if channel signatures are being used. In this research, we looked at availability concerning the emergency vehicle warning. The next section possesses security solutions for all the requirements. In addition to that, some solutions have been provided for the requirements.

7.4.1.5.3 Availability

Availability holds the possibility of being solved implicitly. Numerous methods can develop the availability to a greater extent. Bian et al. (2017) have come up with a solution by installing a channel that can be established to get rid of jamming. The parties can hop between the channels through a particular sequence that is completely unknown to the attacker (see Section 7.2.10). The different solutions include a packet-forwarding approach where the message storms and flooding can be eliminated, initiated by the broadcasts.

7.4.1.5.4 Privacy

Privacy cannot be considered a vital element for the use cases. However, by randomizing the choice based on the base station, it can be solved to some extent, as discussed in Farhang et al. (2015) (see Subsection 7.2.11).

7.4.1.5.5 Confidentiality

Various methods can be appropriate in solving confidentiality by using the PLS (see Subsection 7.2.11). However, in the use cases, confidentiality has no importance. It can be utilized in instant messaging, parking management, or an automatic access control system.

7.4.1.5.6 Authentication

The 5G core network can have an application if the vehicle resides inside the network. The 5G-Authentication and Key Agreement (5G-AKA) protocol (see the case study in Subsection 7.4.3.1) can easily be used for User Equipment (UE), a User Services Identity Module (USIM) that can communicate with the base station. 5G technology has the potential and capability to authenticate users. It is important to find all the necessary information on these technologies so that an efficient and effective one can be selected for V2X communication.

7.4.2 Data analysis of our survey

In the first part, we analyze the primary data that we collected while doing the survey. We conducted the survey, "5G Security Survey," with some internet questions on various social media platforms, such as Facebook, LinkedIn, and so on.

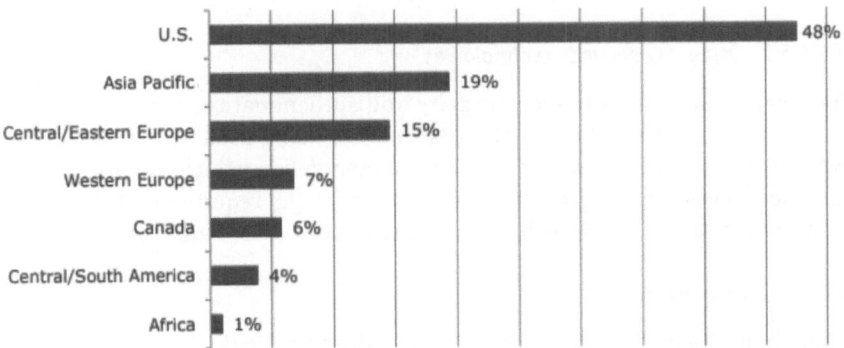

In this survey, we got the above graph's response, which shows that in the United States, people are more concerned about the 5G security than any other country in the world. Africa is the least country concerned and not many people have contributed to the survey.

There was the question, "What might be the concern you will have if you were to emphasize on trial program in term of security proficiencies."

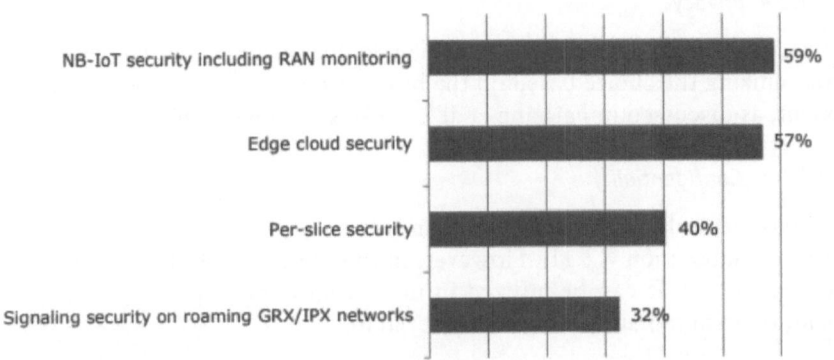

A radio structure that is fully scattered and has perfect control of users' plan by implementing 5G NGC and NR has a major implications.

In response to that, the Customer Service Point (CSP's) had to come up with the solution that can provide fulfilling security measures that consist of all aspects of security; for example, signaling security, business security, and, moreover, IoT and multi-access edge computing (MEC) services.

7.4.2.1 Question: "If you were to launch the 5G on a commercial basis, what would be the architectural design you would follow to support your business?"

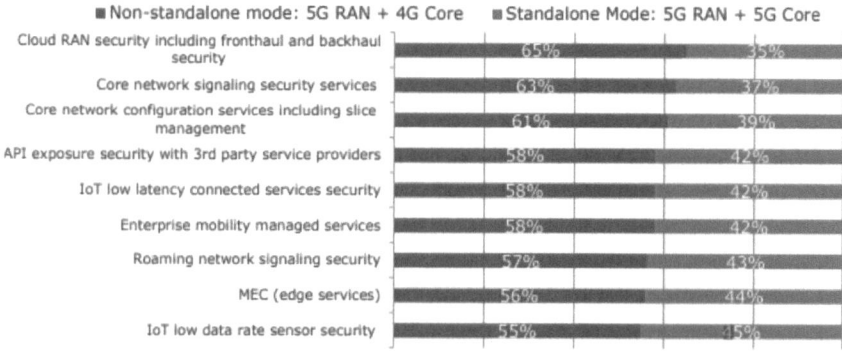

The strong focus is on the cloud radio access network (RAN) security and core network signaling capacity, which attracted 65% and 63%, respectively. It needs to be run in all configurations. This can also simplify the requirement of providing support on the Internetwork Packet Exchange (IPX) so that the main core can have the usage in the case of the launching operation. The National Security Agency of the US (NSA) can easily pave the way for having less complexity with security services.

7.4.2.2 Question: "Can you tell us the preferred encryption method you would choose for securing the user's data in different layers?"

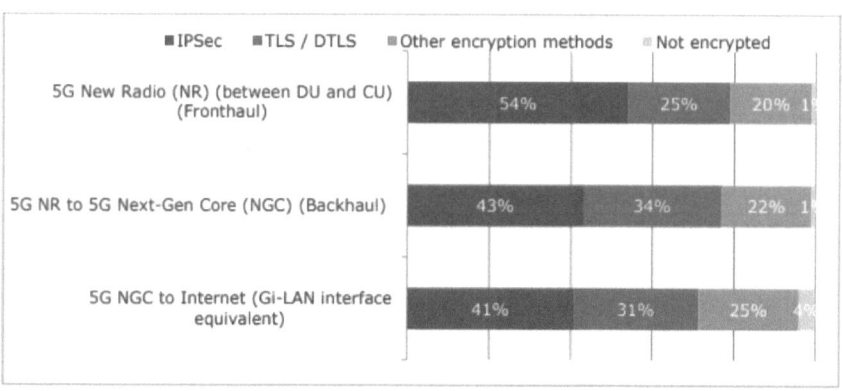

5G is needed to provide support for the encryption in a cloud-based architecture. Hence, it is a must to ace in comprehending encryption preferences throughout different layers of networks for end-to-end security protocols. With a close focus on RAN, the figure (section 7.4.2.2) above presents an impression of the radio network Internet Protocol Security (IPsec) (54%), a type of encryption related to protocol. Additionally, other layers of networks have a preference for the IPsec. However, the support and the responses contributed 43% and 41%, respectively. The usage of the TLS protocol-based approach is the second preference selected by the respondents. The range from 25% to 34% was chosen the core (known as backhaul interfaces) and 31% core, which was done by facing the internet. The Heavy Reading had the support of "other encryption methods" It has a range from 20% RAN to a maximum of 25% RAN concerning the internet-facing interface. Alternative protocol-based approaches can be used in terms of the Quick UDP Internet Connection (QUIC), which is responsible for managing Hypertext Transfer Protocol2 (HTTP/2) services in case of a latent environment.

7.4.2.3 Question: "What would be the timeframe we are looking at for implementing the below-mentioned control panel abilities in the security sector over 5G network."

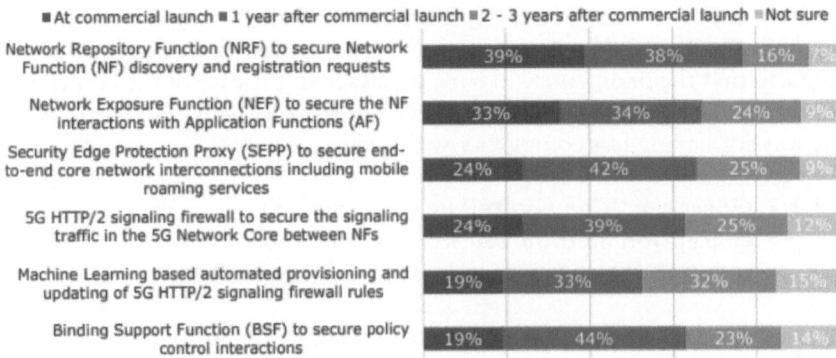

According to the figure above responses, Network Exposure Function (NRF) is highly likely to be executed (39%) to enable service profile delivery. The NEF can provide service as a commercial priority mode (33%). And, it is a tie for third place, with the Security Edge Protection Proxy (SEPP) (24%) for securing roaming and signaling the 5G firewall, which plays an important role in securing signaling operations (24%).

7.4.2.4 Question: "Do you think that it is important to have the existing firewall (for 4G) to keep the 5G network safe in a commercial market structure."

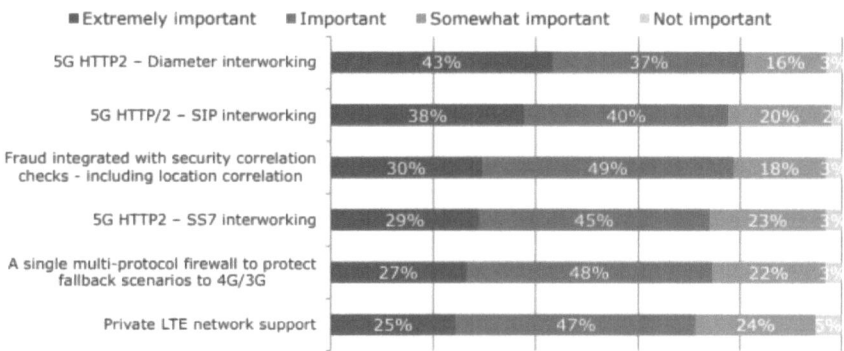

The challenges the CPS control planes face are ones like managing the complex and hybrid environment to ensure the 3G and 4G signaling protocols' have uninterrupted working capacities. There lies an important consideration in controlling the 3G and 4G firewalls, supporting the 5G. Further, all other attributes' "'Extremely Important" inputs relate to HTTP/2: HTTP/2 – Diameter interworking (43%), HTTP2 – Session Initiation Protocol (SIP) interworking (38%), fraud/correlation capabilities (30%), HTTP2 – SS7 interworking (29%), and the single/multi-protocol support firewall (27%), which reflects the need to support HTTP2 Diameter, SIP, and even SS7 interworking. The level of "Important" responses (37% to 49%) is also significant. It reaffirms the focus on fraud/correlation (49%), the single multi-protocol firewall (48%), and even the importance of LTE network support (47%), which continues to gain market traction.

7.4.3 Case studies

The case study is another research technique that compares the past condition with the current condition develop a solution and take cues from what history teaches us. In the next section, we discuss three case studies.

7.4.3.1 Case study one: 5G-AKA and ITS authentication

Arko et al. (2017) published in a paper about security and efficiency. The efficiency of two authentication technologies regarding the ITS model and 5G concerning V2X communication is still up for debate. For moving individuals, these technologies can assist in utilizing limited resources and generating speed. Additionally, the 5G-AKA protocol can be utilized for the users who are not inside the coverage.

On the other hand, the channel response is capable enough to provide fast and smooth authentication. The technology is not suitable for authenticating the sender by identifying the identifier. However, it can provide its services for re-authentication in warning messages that repeat simultaneously. It can be made possible if the authorities require no authentication for identity in delivered messages for audit. Before applying the technology, it is imperative to be able to put trust to make it reliable and verified. Additionally, the verification needs to include the fingerprint along with the moving receivers and senders. Security problems need to be acknowledged and fully checked so that serious caution can be used to protect users from harmful things. Also, beamforming can ensure the improvement of the physical layer authentication. Later on, it can be added to the ITS model. The integration should possess the cryptographic mechanisms and physical layer authentication.

7.4.3.2 Case study two: Adapting IBC for V2X communications

Drias et al. (2017) proposes IBC, which is essential for repudiating the certificates needed in the ITS model. This technology can ease the verification procedure for the keys and is capable of authenticating directly so that no authority needs to be contacted. Suppose IBC can be integrated with the ITS. In this case, it will become one of the most important technologies that can perform better compared to 5G-AKA for V2X. IBC has a lot of promise to bring benefits; however, it needs to adapt to V2X communication. IoT is comparatively a V2X look-alike; hence, the IBC can be suitable for the V2X.

Other modern technologies can enhance new possibilities for the 5G network to make it readily available. CR (Soliman J. et al., 2017) and NOMA (Kizilirmak et al., 2016) scheduling are the two modern technologies that can add value to the 5G. CR scheduling is responsible for efficiently utilizing the resources, whereas NOMA scheduling is responsible for transmitting different signals with identical frequency. If both of the technologies above can be integrated together, network availability will not be an issue. It is recommended to secure the execution procedure given NOMA scheduling has security concerns. The CR possesses a complicated nature that is vulnerable to attackers.

7.4.3.3 Case study three: Ericsson whitepaper

Ericsson (2018) had published a whitepaper on the security of 5G, which states an abysmal view on the usage of various technologies in the 5G arena. The lists include communication security, privacy, resilience, identity management, and security assurance. The communication security for both user plane traffic, as well as signaling, is kept in encryption. Also, the signaling traffic is generally protected. This can be available for user plane traffic. Privacy is a must in this case, and this is only provided by encrypting the identifier. It is a method for protection for the identifier, which is long term in nature. It is also compulsory to refresh the identifier, which is not permanent for short-term purposes only. In the case of resilience, the isolation and separation methods is considered. Identity management plays the same role 4G does, such as securing the cryptographic mechanisms, a system for mutual verification between the user and the network. Additionally, identity management is responsible for the portrait of the Extensible Authentication Protocol (EAP) framework. The EAP framework is the best suited for mobile operators for authentication purposes. The job of security assurance lies in ensuring the network meet the requirements to provide security. This consists of an auditing infrastructure and requirements for security. In recent years, Oxford University published (2018) a report stating the technical vulnerability lies in the 5G-AKA protocol.

7.5 FUTURE DIRECTION AND RECOMMENDATIONS

Now we focus on the challenges that will occur when the establishment takes place between 5G technology and the V2X communication system. It is expected that this research topic will draw a lot of interest and related research. Future recommendations based on the potential research challenges are as follows:

Recommendations

- It is important to find the latency, which is limited to 1 millisecond at a maximum level when emergency situations occur. Further, the automatic pilot, in this case, is very important, and must be on its toes to assist the driver as fast as it can.
- It is important to consider the ultra-reliable communication system concerning the vehicle and the general public's safety. Though much research has been done in this sector, the connection between the IoT and connected devices will surely make the issue a complicated one.
- It is also vital to research the identification of different loopholes that might prove to be vulnerabilities in the communication system. Hence, it is compulsory that a lot of work be put into the physical layer, to make the identification easier for the data and equipment. Another research challenge can be Machine Learning (ML) techniques, which can efficiently analyze the attackers' abnormal behavior.
- In the case of unmanned aerial vehicle (UAV) operations, the main factors are reliability and latency. These factors are needed to ensure that safety measures can be provided in the operating system. It is

also important to have a constant connection between the ground station and UAV for uninterrupted video streaming.

- In UAV communication, there are multiple problems related to privacy and security, which need to be addressed. The problems relate to spoofing, jamming, and eavesdropping. To provide a solution to the problems, it is essential to follow Artificial Intelligence (AI) solutions and light-weight techniques.
- Other challenges related to UAV communication can be the avoidance of collisions and fast and mobile support, which can be researched.
- A reliable connection, along with the high-speed network, is required to monitor processes. If it is not available, researchers will not possess exact, real-time data to make sound decisions. However, getting high speed and a secured connection is a problem that the researchers needed to address.
- It is necessary to support the 3GPP SA3 security protocols and engage with the system. In a given time, it is important to become realistic and pragmatist. In contrast, it is not expected to miss any opportunities to devise new things to improve 5G.
- It is mandatory to provide encouragement to support software-defined networking/network functions virtualization (SDN/NFV) security. In addition, stimulation for the products is required that can provide support to sensitive functionalities in different types of visualized environments.
- It is also advised to have proper communication with open-source communities. Also, advocacy to bring more protection to the open source is needed.
- The assurance of the security-by-default has been built in 5G. It is also suggested by the Fast-Moving Consumer Goods (FCCG) report that security will be needed to be demonstrated as part of the development program. Future research should also consider developers' engagement in 5G test buds' design to establish security and protection in the initial stage.

7.6 CONCLUSION

5G is a recent phenomenon in this world, and it is always making peoples' lives better. Additionally, V2X communication has existed for some time to protect the secrecy of thousands. It has been one of the trusted modes of communication. This chapter contains the study on the effect of 5G security focused on the V2X network, which can impact the security measures of V2X communications. We have discussed related literature and collected participants' responses to analyze the impact of V2X communications on the introduction of 5G technology. In the beginning, we adopted the ETSI ITS model to describe the current status and standards of V2X communications. We covered security procedures and prospects in detail. Later, we presented solutions to the security problems discussed concerning the 5G. The analysis part of this chapter depicts the security requirements in various use cases. In addition, corresponding solutions are presented based on the requirements.

An investigation on the security aspects included different variables germane to the authentication, privacy, confidentiality, availability, and integrity of the system. The research discovered that the 5G-AKA protocol and the physical layer authentication should replace heavy cryptographic algorithms in the ITS given the 5G-AKA protocol is proven to be more efficient than the latter one. IBC needs to be adapted for the V2X communications. Then it is be wise to remove the certificate infrastructure of the ITS model. We discussed future prospects and directions pertinent to this research field. The existence of V2X communications came long before the initiation of 5G. During that time, security mechanisms and the protocol stack have been made for standardization with state-of-the-art technology assistance.

Nevertheless, it cannot be denied that 5G advances towards the business end and possesses modern standards to make lives easier, more comfortable, and more secure. Given 5G possesses technological advancements, many researchers believe that 5G will have a huge grasp on V2X communications. However, it is important to emphasize that 5G security needs to be analyzed thoroughly. It is necessary to find an efficient way to integrate with the existing ITS model.

REFERENCES

5G Automotive Association. The Case for Cellular V2X for Safety and Cooperative Driving. Technical Report 23-Nov-2016, 5G Automotive Association, Neumarkter Str. 21 81673, Munich, Germany, 2016.

5G Automotive Association. An assessment of direct communications technologies for improved road safety in the EU. pages 1–80, December 2017.

5GCAR consortium. 5GCAR first report: The 5GCAR EU initiative pushes for future wireless vehicular communication. 2017. URL https://5gcar.eu/wpcontent/uploads/2017/11/First-5GCAR-Press-release_20171017.pdf.

5GPPP Architecture Working Group. 5GPPP Architecture Working Group View on 5G Architecture. Technical Report Jan-2018-v2.0, 5GPPP, Wieblinger Weg 19/4, 69123 Heidelberg, Germany, December 2017.

Abdel Hakeem, S. A., Hady, A. A., and Kim, H. (2020). Current and future developments to improve 5G-NewRadio performance in vehicle-to-everything communications. *Telecommunication Systems*. doi:10.1007/s11235-020-00704-7

Aguilera T., Alvarez, F. J., Sanchez, A., Albuquerque, D. F., Vieira, J. M.N., and Lopes, S. I. Characterization of the Near-Far problem in a CDMA-based acoustic localization system. Proceedings of the IEEE International Conference on Industrial Technology, 2015-June(June):3404–3411, 2015. doi: 10.1109/ICIT.2015.7125604.

Ahmed, S. A., Ariffin, S. H., and Fisal, N. (2013). Overview of Wireless Access in Vehicular Environment (WAVE) protocols and standards. *Indian journal of science and technology*, 6(7), 4994-5001. Retrieved 7 29, 2019, from http://indjst.org/index.php/indjst/article/viewfile/34355/27974

Al-Momani, A., Kargl, F., Waldschmidt, C., Moser, S., and Slomka, F. Wireless channel-based message authentication. IEEE Vehicular Networking Conference, VNC, pages 271–274, January 2016. ISSN 21579865.doi: 10.1109/VNC.2015.7385587.

Ali, A., Cao, H., Eichinger, J., Gangakhedkar, S., and Gharba, M. A Testbed for Experimenting 5G-V2X Requiring Ultra Reliability and Low-Latency. WSA 2017; 21th International ITG Workshop on Smart Antennas, pages 1–4, 2017.

Anggorojati, B. and Prasad, R. Securing Communication in Inter Domains Internet of Things using Identity-based Cryptography. 2017 International Workshop on Big Data and Information Security (IWBIS), pages 137–142, 2017.

Aurora, R. N., Zak, R., Auerbach, S., Casey, K. R., Chowdhuri, S., Karippot, A., ... Tracy, S. L. (2010). "Best practice guide for the treatment of nightmare disorder in adults." *Journal of Clinical Sleep Medicine*, 6(4), 389–401. Retrieved 7 29, 2019, from https://ncbi.nlm.nih.gov/pmc/articles/pmc2919672

Bernard, H. (2011). Research Methods in Anthropology. In *Research Methods in Anthropology* (p. 7). AltaMira Press.

Bian, K., Zhang, G., and Song, L. Toward Secure Crowd Sensing in Vehicle-to-Everything Networks. *IEEE Network*, pages 1–6, 2017. ISSN 08908044. doi: 10.1109/MNET.2017.1700098.

Boban, M., Manolakis, K., Ibrahim, M., Bazzi, S., and Xu, W. Design aspects for 5G V2X physical layer. 2016 IEEE Conference on Standards for Communications and Networking, CSCN 2016, 2016. doi:10.1109/CSCN.2016.7785161.

Drias, Z., Serhrouchni, A., and Vogel, O. (2017). Identity-Based Cryptography (IBC) Based Key Management System (KMS) for Industrial Control Systems (ICS). *Cyber Security in Networking Conference (CSNet)*, 2017 1st, pages 1–10.

European Telecommunications Standards Institute. Network domain security; authentication framework; (release 6). ETSI Technical Specification Group Service and System Aspects TS 133 310, 3GPP, 650 Route des Lucioles F06921 Sophia Antipolis Cedex - FRANCE, February 2004.

European Telecommunications Standards Institute. 3G Security; Network Domain Security; IP network layer security (Release 9). ETSI Technical Specification Group Service and System Aspects TS 133 210, 3GPP, 650 Route des Lucioles F-06921 Sophia Antipolis Cedex - FRANCE, 2009.

European Telecommunications Standards Institute. Basic Set of Applications; Definitions. ETSI technical report on Vehicular Communications TR 102 638 - V1.1.1, ETSI, 650 Route des Lucioles F-06921 Sophia Antipolis Cedex - FRANCE, 2009.

European Telecommunications Standards Institute. Part 3: Specifications of Decentralized Environmental Notification Basic Service. ETSI technical specification on Vehicular Communications TS 102 637-3, ETSI, 650 Route des Lucioles F-06921 Sophia Antipolis Cedex - FRANCE, 2010. 50 Bibliograph

European Telecommunications Standards Institute. Security Services and Architecture. ETSI technical specification on Intelligent Transport Systems (ITS) TS 102 731, ETSI, 650 Route des Lucioles F-06921 Sophia Antipolis Cedex - FRANCE, 2010.

European Telecommunications Standards Institute. Communications Architecture. ETSI european standard on Intelligent Transport Systems (ITS) EN 302 665 - V1.1.1, ETSI, 650 Route des Lucioles F-06921 Sophia Antipolis Cedex - FRANCE, 2010.

European Telecommunications Standards Institute. Part 5: Transport Protocols; Sub-part 1: Basic Transport Protocol. ETSI technical specification on Vehicular Communications TS 102 636-5-1 - V1.1.1, ETSI, 650 Route des Lucioles F-06921 Sophia Antipolis Cedex - FRANCE, 2011

European Telecommunications Standards Institute. Basic Set of Applications; Part 2: Specification of Cooperative Awareness Basic Service. ETSI technical specification on Vehicular Communications TS 102 637-2, ETSI, 650 Route des Lucioles F-06921 Sophia Antipolis Cedex - FRANCE, 2011.

European Telecommunications Standards Institute. ITS communications security architecture and security management. ETSI technical specification on Intelligent Transport Systems (ITS) TS 102 940 - V1.1.1, ETSI, 650 Route des Lucioles F-06921 Sophia Antipolis Cedex - FRANCE, 2012.

European Telecommunications Standards Institute. Security; Trust and Privacy Management. ETSI technical specification on Intelligent Transport Systems (ITS) 102 941 - V1.1.1, ETSI, 650 Route des Lucioles F-06921 Sophia Antipolis Cedex - FRANCE, 2012.

European Telecommunications Standards Institute. Security; Security header and certificate formats. ETSI technical specification on Intelligent Transport Systems (ITS) TS 103 097 - V1.1.1, ETSI, 650 Route des Lucioles F-06921 Sophia Antipolis Cedex - FRANCE, 2013.

European Telecommunications Standards Institute. Universal Mobile Telecommunications System (UMTS); LTE; 3GPP System Architecture Evolution (SAE); Security architecture. ETSI Technical Specification on Digital cellular telecommunications system (Phase 2+) TS 133 401 - V10.3.0, ETSI, 650 Route des Lucioles F-06921 Sophia Antipolis Cedex - FRANCE, 2013.

European Telecommunications Standards Institute. Service requirements for V2X services. ETSI technical specification on LTE TS 122 185 - V14.3.0 Release 14, ETSI, 650 Route des Lucioles F-06921 Sophia Antipolis Cedex - FRANCE, 2017.

European Telecommunications Standards Institute. Architecture enhancements for V2X services. ETSI Technical Specification on Universal Mobile Telecommunications System (UMTS); LTE TS 123 285 - V14.3.0, ETSI, 650 Route des Lucioles F-06921 Sophia Antipolis Cedex - FRANCE, 2017.

European Telecommunications Standards Institute. Proximity-based services (ProSe); Stage 2. ETSI Technical Specification on Universal Mobile Telecommunications System (UMTS); LTE TS 123 303 - V14.1.0, ETSI, 650 Route des Lucioles F-06921 Sophia Antipolis Cedex - FRANCE, 2017.

European Telecommunications Standards Institute. Security aspect for LTE support of Vehicle-to-Everything (V2X) services. ETSI Technical Specification on LTE; 5G TS 133 185 - V14.0.0, ETSI, 650 Route des Lucioles F-06921 Sophia Antipolis Cedex - FRANCE, 2017.

Farhang, S., Hayel, Y., and Zhu, Q. PHY-layer location privacy-preserving access point selection mechanism in next-generation wireless networks. 2015 IEEE Conference on Communications and NetworkSecurity, CNS 2015, pages 263–271, 2015. doi: 10.1109/CNS.2015.7346836.

Gianotti, F., Tacchini, A., Leto, G., Martinetti, E., Bruno, P., Bellassai, G., ... Trifoglio, M. (2016). Information and Communications Technology (ICT) Infrastructure for the ASTRI SST-2M telescope prototype for the Cherenkov Telescope Array. Proceedings of SPIE, 9913. Retrieved 7 29, 2019, from https://spiedigitallibrary.org/conference-proceedings-of-spie/9913/1/information-and-communications-technology-ict-infrastructure-for-the-astri-sst/10.1117/12.2230150.full

Haas. H., Yin, L., Chen, C., Videv, S., Parol, D., Poves, E., Alshaer, H., and Islim, M. (2020). "Introduction to indoor networking concepts and challenges in LiFi," *J. Opt. Commun. Netw.* 12, A190–A203.

Karthik, P., Muthu Kumar, B., Ravikiran, B.A., Suresh, K., and Toney, G. Implementation of visible light communication (VLC) for vehicles. Proceedings of 2016 International Conference on Advanced Communication Control and Computing Technologies, ICACCCT 2016, (978):673–675, 2017. doi: 10.1109/ICACCCT.2016.7831724.

Kizilirmak, R. C,. Non-Orthogonal Multiple Access (NOMA) for 5G Networks. In Hossein Khaleghi Bizaki, editor, *Towards 5G Wireless Networks - A Physical Layer Perspective*, chapter 04. InTech, Rijeka, 2016. doi: 10.5772/66048. URL http://dx.doi.org/10.5772/66048.

Larsson, K., Halvarsson, B., Singh, D., Chana, R., and Manssour, J. High-Speed Beam Tracking Demonstrated Using a 28 GHz 5G Trial System. Vehicular Technology Conference (VTC-Fall), 2017 IEEE 86th, 2017.

Luoto P., Bennis, M., Pirinen, P., Samarakoon, S., Horneman, K., and Latva-Aho, M. Vehicle clustering for improving enhanced LTEV2X network performance. EuCNC 2017 - European Conference on Networks and Communications, pages 1–5, 2017. doi: 10.1109/EuCNC.2017.7980735.

McCusker, K., and Gunaydin, S. (2015). Research using qualitative, quantitative or mixed methods and choice based on the research. *Perfusion- SAGE*, 30(7), 537–542.

Pak, W. (2017). Fast packet classification for V2X services in 5G networks. *Journal of Communications and Networks*, (3):218–226. ISSN 12292370. doi:10.1109/JCN.2017.000039.

Pan, F., Wen, H., Song, H., Jie, T., and Wang, L. 5G Security Architecture and Light Weight Security Authentication. 2015 IEEE/CIC International Conference on Communications in China: First International Workshop on Green and Secure Communications Technology, 2015.

Satrya, G. B. and Shin, S. Y. Security Enhancement to Successive Interference Cancellation Algorithm for Non-Orthogonal Multiple Access 53 Bibliography (NOMA). 2017 IEEE 28th Annual International Symposium on Personal, Indoor, and Mobile Radio Communications (PIMRC), pages 5–9, 2017.

Saunders, M., Thornhill, A., and Lewis, P. (2009). Understanding research philosophies and approaches. In Pearson (Ed.), *Research methods for business students* (pp. 106–135). Essex: Pitman Publishing.

Soliman, J. N. and Mageed, T.A. Taxonomy of Security Attacks and Threats in Cognitive Radio Networks. 2017 Japan-Africa Conference on Electronics, Communications and Computers (JAC-ECC), pages 127–131, 2017 networks: A Review. China Communications, pages 1–14, 2017.

Tateishi, K., Kurita, D., Harada, A., Kishiyama, Y., Itoh, S., Murai, H., Schrammar, N., Simonsson, A., and Okvist, P. Experimental evaluation of advanced beam tracking with CSI acquisition for 5G radio access. IEEE International Conference on Communications, 2017. ISSN 15503607. doi: 10.1109/ICC.2017.7996953.

Visala, K. (2014). Hybrid Communication Architecture HCA. arXiv: Distributed, Parallel, and Cluster Computing. Retrieved 7 29, 2019, from https://arxiv.org/pdf/1407.4149.pdf

Yu R., Bai Z., Yang, L., Wang, P., Move, O. A., and Liu, Y. (2016). A Location Cloaking Algorithm Based on Combinatorial Optimization for Location-Based Services in 5G Networks. *IEEE Access*, 4:6515–6527. ISSN 21693536. doi: 10.1109/ACCESS.2016.2607766.

Zhou, C. and Kellerer, W. Multi-User-Centric Virtual Cell Operation for V2X Communications in 5G Networks. 2017 IEEE Conference on Standards for Communications and Networking (CSCN), pages 84–90, 2017.

Chapter 8

Security analysis for VANET-based accident warning systems[*]
BOOK CHAPTER

Niaz Chowdhury[†]
Knowledge Media Institute, The Open University
Milton Keynes, United Kingdom

Lewis Mackenzie
Department of Computing Science University of Glasgow
Glasgow, United Kingdom

CONTENTS

[*] This preprint will be published as a book chapter (Chapter 8) in *Vehicular Communications for Smart Cars: Protocols, Applications and Security Concerns* by Taylor & Francis.
[†] This work is funded by the Scottish Overseas Research Students Award (SORSA) and the University Glasgow College of Science and Engineering Scholarship.

DOI: 10.1201/9781315110905-8

8.1 INTRODUCTION

Accident warning systems (AWSs) are developed for next-generation vehicles, and aim to use vehicular ad-hoc networks (VANETs) to avoid potential collisions and spread safety notifications amongst nearby vehicles [1, 2]. The problem of designing efficient and effective warning systems has been widely studied but making such systems secure from potential threats has yet to be seriously addressed.

Although security is considered an important issues in general networking, it has largely been overlooked in warning systems developed for vehicles. There is sometimes an implicit assumption that a separate security system designed for generic wireless networks can be added to an AWS [3]; however, such an approach is unlikely to be adequate because of the unique nature of both the safety system itself and the potential threats. For example, unlike most VANET applications that often handle confidential and sensitive data, warning systems are not generally concerned about data confidentiality. These systems willingly share data with other nodes so that they can operate cooperatively to prevent motor accidents. The special nature of AWSs makes it necessary to develop a specific threat model by anticipating potential adversaries, their motivations, and likely modes of attack. This chapter describes and develops a threat model through an in-depth analysis of security, trust, and privacy issues in AWSs, and addresses future ways to contribute in this area. The offerings of this chapter are presented twofold: first, it presents a survey of possible adversaries and potential attacks on AWSs; and, second, it develops a threat model by ranking adversaries based on the level of potential damage associated with their likely types of intrusion (Figure 8.1).

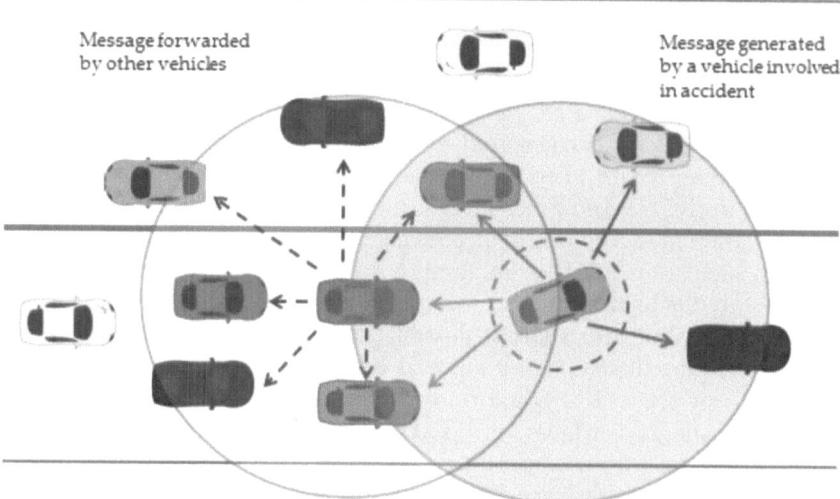

Message forwarded
by other vehicles

Message generated
by a vehicle involved
in accident

Figure 8.1 The operation of an AWS in brief. The remainder of the chapter is structured
as follows: Section 8.2 presents an overview of the system; Sections 8.3 and
8.4 discuss adversaries and attacks, respectively; Section 8.5 outlines potential challenges; Section 8.6 presents the threat model; and, finally, Section 8.7
concludes the chapter with a summary.

8.2 OVERVIEW OF THE SYSTEM

A threat model's design process requires a preliminary requirement analysis
and component-level description of the system. Such a study is presented in the
publication [4]. We summarize the main conclusions of that study briefly next.

8.2.1 System architecture and requirements

AWSs are a collections of mobile nodes, each corresponding to a physical
vehicle, whose purpose is to generate collision avoidance notifications that
warn drivers before a potential accident occurs. These systems operate via
a VANET with which they may be integrated to a greater or lesser degree.

With system designs, collisions may be classified into several generic
types: follow-up, pile-up, intersection, lane-change, forward-collision, and
collisions with an object, human, or animal. An AWS will use its associated VANET to send messages containing information about vehicles'
locations and velocities to tackle these various hazards. To generate such
information, individual vehicles need to be equipped with various devices,
including Global Positioning System (GPS) or similar receivers, sensors to
gather important data such as speed, acceleration, and deceleration, and
the presence of other vehicles nearby, and possibly On-Board Units (OBUs)
to allow drivers to enter warnings of less time-critical hazards manually.

8.2.2 Message pattern

Information that vehicles exchange that is not confidential is visible to, and accessible by, everyone. This information can be used to identify potential hazards to as wide a relevant audience as possible. There are generally four types of AWS messages used to disseminate information: the Event-Driven Message (EDM), Period Warning Message (PWM), Road Condition Notification (RCN), and Emergency Call-Up (ECU).

EDMs are sent in response to an emergency that arises suddenly and unexpectedly, such as an accident, an abruptly stopped vehicle, and so on. These messages are the highest priority variants in any AWS and need to reach the targeted audience in the shortest possible time. PWMs are also a high priority and require dissemination quickly; vehicles send these to warn others about their presence. RCNs inform other vehicles about scenarios such as damaged or slippery surfaces, localized weather hazards, and so on, and are treated as low-priority notifications. ECU is used in some AWSs to summon the police, ambulances, or mechanics after an incident takes place.

8.2.3 Interactions

AWS messages are typically small and open in nature. Senders want every recipient to which the information might be relevant to read the messages, and messages are forwarded repeatedly over a specific region. Unfortunately, this can allow attackers to manipulate content and spread false information across the network. By its nature, in an AWS that depends on accurate content, manipulation introduces not only the threat of the system failing to work as intended but also, even worse, the possibility of it actually causing accidents that would otherwise not have occurred.

8.3 ADVERSARIES

Potential attackers can be categorized according to the damage they might cause. The following section divides them into three classes and describes how they are likely to operate in AWSs (see Figure 8.2).

8.3.1 First-degree threat

This category includes adversaries whose objective primarily involves breaching normal practices to temporary personal driving gains. The aim is not to cause physical harm to others or result in a monetary gain.

8.3.1.1 Dishonest drivers

Dishonest drivers are potential adversaries who inject false information into the network to gain an advantage over other vehicles. For example,

Figure 8.2 List of adversaries based on their degree of threat.

one might create the illusion of congestion to encourage other drivers to avoid a route one wishes to use freely.

8.3.1.2 Selfish drivers

AWSs are cooperative system, and every vehicle must comply with this principle. To make the system a success, it is assumed that all vehicles will share information and forward it to others, if necessary. It is, however, possible that some drivers may refuse to comply with this norm by disabling the forwarding of warning messages, either completely or partially. This behavior's impact may or may not cause a system failure depending on its prevalence in a particular area.

8.3.1.3 Pranksters

These are likely to be amateur hackers interfering with the system for amusement and, possibly, notoriety. Attacks might be carried out from the roadside, where hackers deliberately feed misleading instructions to vehicles. Most such individuals intend on creating an inconvenience rather than serious damage, but given the nature of motor vehicles, there is a danger of unintended serious harm.

8.3.1.4 On-board passengers

Warning systems are often equipped with an OBU that helps drivers entering warning and road-condition notifications manually. Although the intention is to allow drivers to report potential hazards where data entry is insecure, it may be possible for passengers to inject false notification into the network, either through carelessness or for amusement.

8.3.2 Second-degree threat

This category comprises adversaries that deliberately attack the system to achieve monetary benefits but do not intend to cause physical harm. This includes misusing the system to rob others or trying to get personal information with to the intention of selling it.

8.3.2.1 Robbers

AWSs warn other vehicles about potential collisions. In an unsecured system where vehicles react automatically to such warnings, it would be possible to inject false information to bring a target vehicle to a halt to facilitate a robbery.

8.3.2.2 Sniffers

Although the warnings that the systems disseminate are public and do not contain any personal information, monitoring a target individual's movement, driving pattern, and vehicle information without consent is a privacy violation and could be used against the victim.

8.3.2.3 Industrial competitors

The Media Access Control (MAC) address of IEEE 802.11 contains a manufacturer identity field. An intruder could, in principle, use this to defame competitors by falsely associating deliberately engineered problems with their systems.

8.3.2.4 Watchers

Watchers encompasses everyone from a government secret service operative to a tabloid newspaper journalist to a criminal group attempting to monitor someone's regular movements and activities. The adversaries' nature is different from sniffers as they explicitly try to defame or watch someone for their monetary or personal gain.

8.3.3 Third-degree threat

The third-degree threat category comprises adversaries whose intention is to interfere with an AWS deliberately to cause harm to others.

8.3.3.1 Malicious attackers

These attackers insert malicious information into the network or jam network channels to block information propagation in the network to create chaos for a variety of potential reasons, such as affecting markets and creating diversions. An attack like this could easily lead to fatal accidents.

8.3.3.2 Terrorist/activists

Terrorists and activists are a group of adversaries that can potentially manipulate data to create fatal accidents on motorways or in crowded city areas.

8.3.3.3 Abductors

An individual, or a group, who wants to abduct someone can take advantage of this system. Drivers inject data regularly into the network, particularly in the periodic warning message, leaving traces on the driver's regular movement path. Abductors can use those traces to predict someone's movement in advance to plot a physical attack in a planned and organized way.

8.3.3.4 Crazy pranksters

For reasons previously mentioned, people pranking an AWS can cause physical injury or death, even if it is done unintentionally. Unfortunately, there is reason to believe that some such individuals are capable of crossing the line and generating dangerous outcomes deliberately for nihilistic pleasure.

8.4 ATTACKS

In this section, we present a comprehensive survey that addresses various types of attack that can target AWSs (see Figure 8.3). Later in this chapter, we connect these attacks and previously described adversaries by building a threat model.

8.4.1 Distributed denial-of-service (DDoS) attack

DDoS is a type of denial-of-service (DoS) attack on networks that is triggered by first compromising several *slave* or *zombie* devices and later using a trigger command to use their combined transmission power in an orchestrated flooding attack on some selected target [5]. The distributed nature of the attack makes it more difficult for the victim to block it. Attacks of this type can make services unavailable during the course of the assault. The DDoS attack is considered one of the most dangerous attacks on networks [6].

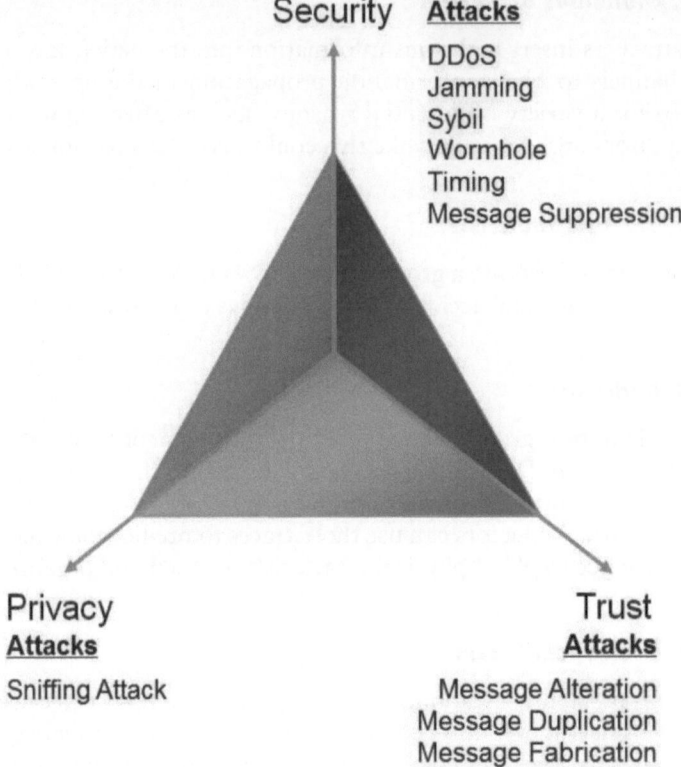

Figure 8.3 Possible attacks associated with security, privacy and trust.

8.4.2 Jamming attack

Wireless networks are built on a shared medium that allows potential adversaries to launch attacks easily. A jamming attack is a DoS attack aimed at disabling this medium. For example, an attacker may use a jamming device that emits a powerful radio frequency (RF) signal to block a wireless channel so that legitimate users cannot access the medium [7, 8]. There are several jamming attack models, such as constant jammer, deceptive jammer, dom jammer, reactive jammer and so on [9]. In an AWS, an attack of this type would make vehicle information unavailable to other nodes. In a scenario where vehicles come to rely on AWSs to trigger automatic avoidance measures, this could easily result in collisions and serious harm.

8.4.3 Sybil attack

The Sybil attack mode was first formally defined by [10] as sending messages from one node with multiple spoofed identities. In a VANET ([11]), this amounts to a vehicle maliciously fabricating different identities to mislead others and generate false information. It is observed and argued that this attack is always possible without a centralized authority and may go undetected [10, 12].

8.4.4 Wormhole attack

The wormhole is an attack on various types of networks and considered one of the most difficult to counter [13]. According to [14], in a wormhole attack, a private tunnel is used to deliver packets that appear to originate locally to a remote destination. In this way, a distant node can be made to appear at a location where it does not exist in reality. A wormhole attack can be performed even if the network provides confidentiality and authenticity.

8.4.5 Sniffing attack

The aim with a sniffing attack is simply to snoop on a targeted individual by collecting private information from the network [15]. There are several variants, including content sniffing, phishing, location sniffing, identity sniffing, and so on. Sniffing attacks have the potential to create a great threat on lives and properties should victims' privacy is compromised.

8.4.6 Timing attack

This attack particularly targets real-time applications. The attacker interferes with a legitimate message to engineer a deliberate delay [16]. Real-time applications that are time dependent can then be made to fail. This kind of attack is tough to detect as the attacker acts like a normal node.

8.4.7 Message attacks

Message attacks are a group of attacks involving message alteration, duplication, fabrication, and suppression. These have been identified as threats for VANETs in studies [17] and target the relaying networks. In alteration, important information is altered during a relay but in such a way that it still looks legitimate. In duplication, a message is replicated by a relay to gain specific objectives. In fabrication, a message is generated that looks like it has been legitimately relayed rather than sourced by its creator. Finally, in suppression, a relay discards a message it is supposed to forward, allowing an attacker to block information from reaching intended recipients.

8.5 POTENTIAL CHALLENGES

AWS differ from other applications that might be expected to run on top of VANETs. They exhibit open behavior where data must be visible, and information about physical locations needs to be exchanged [18, 19]. Because of this special nature, defending against threats presents some important challenges.

8.5.1 Cooperative systems

By its nature, an AWS is a cooperative distributed system operating locally with no centralized control. The basic conception relies on every node acting in an honest, helpful, cooperative manner because each is reliant for its safety on the received data. However, in practice, adversaries can prey on a system organized to achieve unethical and illegal benefits. It is easy to trigger attacks by using suitably prepared intruder vehicles to inject false or fabricated data into the network. A potential solution to this problem would be using an existing trust system, but that may create tension with a desire to preserve drivers' personal privacy.

8.5.2 Trust vs. privacy

Trust and privacy always have a somewhat uneasy relationship that becomes especially complicated in AWSs. Nodes depend on received data for important and safety-critical decision-making, and they need to know that such data is coming from a legitimate source. While a trust mechanism, by endorsing data, would remove many potential threats, in this situation, given inherent location tracking, it would also compromise privacy and open drivers to snooping and even criminal targeting.

8.5.3 Non-confidential system

In an AWS, the aim is to disseminate data to all vehicles affected by the content. As a result, the source is not usually aware of the relevant audience and cannot use secure channels. This makes it easier for attackers to alter or fabricate data and disseminate false information that looks legitimate.

8.5.4 Decentralized nature

The openness issue might be addressed if vehicles are provided with certificates issued by a trusted central authority. In this case, only messages signed by their sources are to be trusted. However, the biggest challenge to doing this is the decentralized nature of AWS, which may not have access to a backbone network and, hence, a central server.

Upon consideration, it is apparent that the challenges discussed previously are interconnected in nature. As decentralized AWSs are a cooperative system, they need to verify trust. Verifying trust, however, affects personal privacy that can be compromised because of the openness of the system.

Nonetheless, openness is difficult to address unless the system is turned into a central authority-controlled system.

8.6 THE THREAT MODEL

Potential adversaries and possible attacks have already been discussed, along with different challenges that AWSs create because of their openness and cooperative behavior. These considerations provide sufficient context for designing a realistic generic threat model for an AWS. Figure 8.4 combines that information, identifying relationships between adversaries and attacks, and presents the threat model and a discussion on a possible countermeasure to safeguard the system from those attacks.

The threat model is built in such a way that it divides the adversaries in the first place. This division provides a clear understanding of the possible attackers who retain the maximum potential to make the system vulnerable. It also shows who often go undetected without leaving any trace of the system being misused. These adversaries can trigger attacks targeting security, privacy, and trust aspects of the system that have also been shown in the model.

To most suitable way to protect AWSs from these attacks is to look at the attack based on the features that are being targeted in the system. For example, dishonest-drivers and pranksters would likely use data-related attacks because the main objective is to fool people to gain a temporary personal benefit or pleasure. Robbers and abductors would try to stop people in the middle of their routes by providing false warnings, while terrorists and malicious attackers would create chaos by injecting malicious information. These behaviors indicate that the attacks would possibly take advantage of trust issues. Therefore, to protect AWSs from such attacks, a solution is establishing a trust system [16].

A number of approaches help establish trust, such as key-based approaches [20], reputation-based approaches [21], and so on. These approaches, however, require identifying the person involved in communication. As ASWs willingly and openly disseminate their location, revealing identities can cause severe threats. A potential solution to this problem would be using a privacy-friendly trust system that recently has become a prime research topic in trust [22].

However, a trust system is of little help in combating warning suppression attacks by both dishonest drivers and selfish drivers, and sometimes crazy pranksters. The selfish-node detection mechanism of mobile ad hoc networks (MANETs) can be introduced in VANET to fight against these adversaries [23]. Further, security enforcement through monitoring vehicle communication behavior offers a partial solution to this problem. Another useful corrective measure that might help combat some threats would be OBUs that allow drivers to counter information discovered to

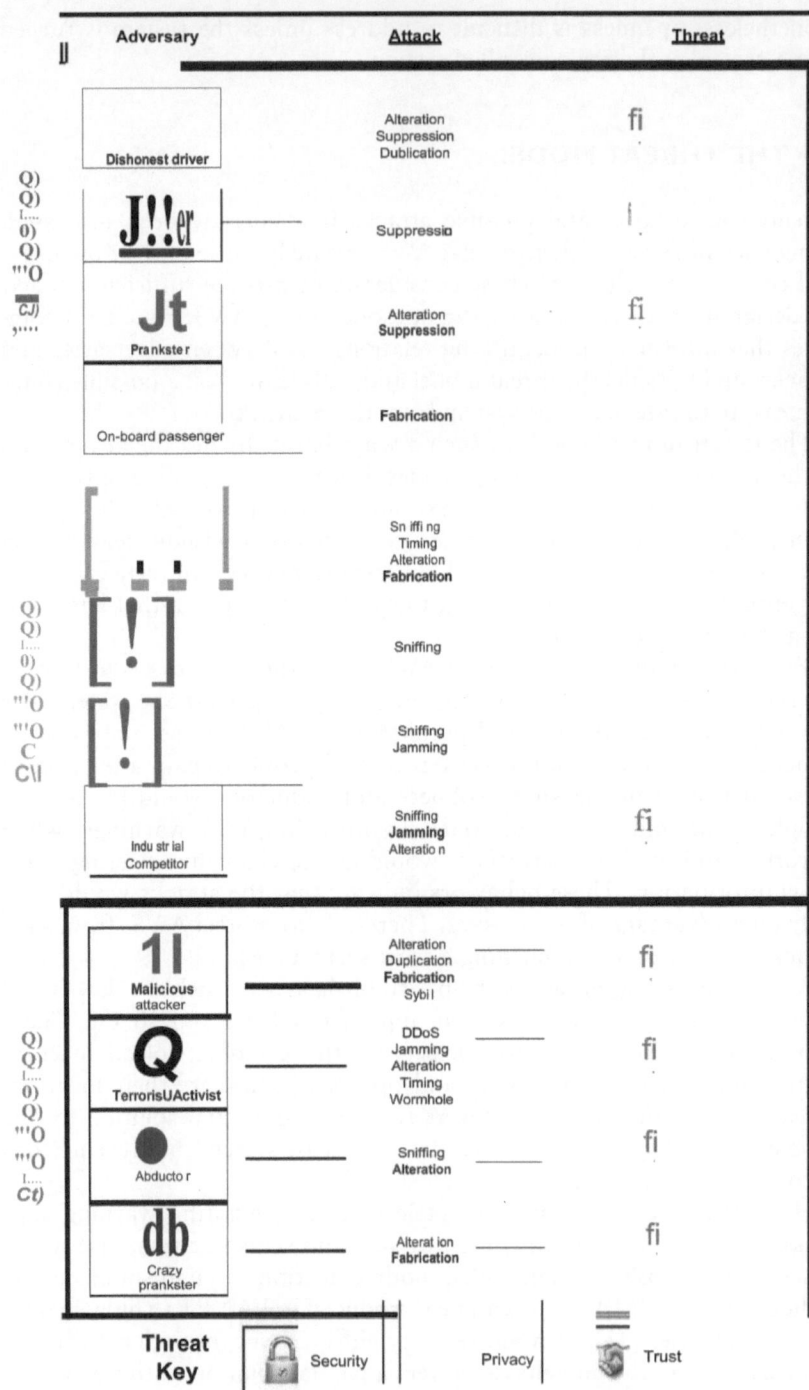

Figure 8.4 The threat model.

be false. This might also help protect against long-term message alterations and cancel out accidental warnings.

8.7 CONCLUSION

This chapter presented an analysis of the security, privacy, and trust issues of the next-generation vehicles, followed by a description of a threat model for warning systems. This chapter's main focus is identifying potential adversaries that might make warning systems vulnerable, and categorizing such adversaries based on their degree of threat and potential attacks they might mount to achieve their objectives. In future, this model can be used to show the right paths in building trust systems and privacy-friendly routing in newly developed warning systems using several new technologies, such as Blockchain [24, 25], Machine Learning [26], the Internet of Things [27] and Decentralized data storage [28, 29].

REFERENCES

[1] Niaz Morshed Chowdhury. *NETCODE: An XOR-based Warning Dissemination Scheme for Vehicular Wireless Networks*. PhD thesis, University of Glasgow, Glasgow, United Kingdom, 2016.

[2] Saif Al-Sultan, Moath M. Al-Doori, Ali H. Al-Bayatti, and Hussien Zedan. A Comprehensive Survey on Vehicular Ad Hoc Network. *Journal of Network and Computer Applications*, pages 1–13, 2013.

[3] Panagiotis Papadimitratos, Levente Buttyan, Tamas Holczer, and Schoch E. Secure Vehicular Communication Systems: Design and Architecture. *IEEE Communications*, 46:100–109, 2008.

[4] Niaz Morshed Chowdhury, Lewis M. Mackenzie, and Colin Perkins. Requirement Analysis for Building Practical Accident Warning Systems based on Vehicular Ad-hoc Networks. In *Proceedings of the 11th IEEE/ IFIP Annual Conference on Wireless On-demand Network Systems and Services (WONS)*, pages 81–88, Obergurgl, Austria, April 2014. IEEE.

[5] Vern Paxson. An Analysis of Using Reflectors for Distributed Denial-of-Service Attacks. *Computer Communica- tion Review*, 31 (3), July 2001.

[6] Lee Garber. Denial-of-Service Attacks Rip the Internet. *Computer Magazine*, July 2000.

[7] Wenyuan Xu, Timothy Wood, Wade Trappe, and Yanyong Zhang. Channel surfing and spatial retreats: defenses against wireless denial of service. In *Proceedings of ACM workshop on Wireless Security*, pages 80–89, 2004.

[8] Anthony D. Wood, John A. Stankovic, and Sang H. Son. JAM: A Jammed-Area Mapping Service for Sensor Networks. In *Proceedings of the 24th Real-Time Systems Symposium*, pages 286–297, Cancun, Mexico, December 2003. IEEE.

[9] Konstantinos Pelechrinis, Marios Iliofotou, and Srikanth V. Krishnamurthy. Denial of Service Attacks in Wireless Networks: The Case of Jammers. *IEEE Communications Surveys and Tutorials*, 13 (2):245–257, 2011.

[10] John R. Douceur. The Sybil Attack. In *Proceedings of the first International Workshop on Peer-to-Peer Systems*, pages 251–260, Berlin, Germany, March 2002. Springer-Verlag.

[11] Chen Chen, Xin Wang, Weili Han, and Binyu Zang. A Robust Detection of the Sybil Attack in Urban VANETs. In *Proceedings of the 29th IEEE International Conference on Distributed Computing Systems Workshops*, Montreal, QC, Canada, June 2009. IEEE.

[12] Gilles Guette and Bertrand Ducourthial. On the Sybil attack detection in VANET. In *Proceedings of IEEE International Conference on Mobile Adhoc and Sensor Systems*, 2007.

[13] Seyed Mohammad Safi, Ali Movaghar, and Misagh Mohammadizadeh. A Novel Approach for Avoiding Wormhole Attacks in VANET. In *Proceedings of First Asian Himalayas International Conference on Internet*, 2009.

[14] Yih-Chun Hu and David Johnson. Wormhole Attacks in Wireless Networks. *IEEE Journal on Selected Areas in Communications*, 24 (2):370–380, 2006.

[15] Bhupendra Singh Thakur and Sapna Chaudhary. Content Sniffing Attack Detection in Client and Server Side: A Survey. *International Journal of Advanced Computer Research*, 2 (10):7–10, 2013.

[16] Irshad Ahmed Sumra, Halabi Hasbullah, Jamalul lail, and Rehman Masood-ur. Trust and Trusted Computing in VANET. *Computer Science Journal*, 1 (1):29–51, 2011.

[17] Bryan Parno and Adrian Perrig. Challenges in Securing Vehicular Networks. In *In the proceedings of ACM SIGCOMM*, 2005.

[18] T. Nadeem, P. Shankar, and L. Iftode. A Comparative Study of Data Dissemination Models for VANETs. In *Proceedings of IEEE Third Annual International Conference on Mobile and Ubiquitous Systems: Networking & Services*, pages 1–10, San Jose, CA, USA, 2006.

[19] Francisco J. Martinez, Juan-Carlos Cano, Carlos T. Calafate, and Pietro Manzoni. A VANET Solution to Prevent Car Accident. In *Proceedings of Jornadas de Paralelismo*, Spain, 2007.

[20] Jian-Jun Wang, Jian-Ping Li, Yong-Fan Li, and Jing Peng. Review of Key-Based Dynamic Trust Authorization Mechanism. In *Proceedings of International Conference on Wavelet Active Media Technology and Information Processing (ICWAMTIP)*, pages 263–267, Chengdu, China, 2012.

[21] Amira Bradai, Walid Ben-Ameur, and Hossam Afifi. Byzantine resistant reputation-based trust management. In *Proceedings of the 9th International Conference on Collaborative Computing: Networking, Applications and Worksharing*, pages 269–278, Austin, TX, USA, October 2013. IEEE.

[22] Sebastian Ries, Marc Fischlin, Leonardo A. Martucci, and Max Muhlhauser. Learning Whom to Trust in a Privacy-Friendly Way. In *Proceedings of 10th International Conference on Trust, Security and Privacy in Computing and Communications (TrustCom)*, pages 214–225, Changsha, China, 2011.

[23] Hongmei Deng, R. Xu, J. Li, and F. Zhang. Agent-based cooperative anomaly detection for wireless ad hoc networks. In *Proceedings of the 12th International Conference on Parallel and Distributed Systems*, Minneapolis, MN, USA, July 2006. IEEE.

[24] Niaz Chowdhury. *Inside Blockchain, Bitcoin, and Cryptocurrencies*. Taylor & Francis, 1st edition, 2019.

[25] Md Sadek Ferdous, Mohammad Jabed Morshed Chowdhury, Kamanashis Biswas, Niaz Chowdhury, and Vallipuram Muthukkumarasamy. Immutable Autobiography of Smart Cars Leveraging Blockchain Technology. *Knowledge Engineering Review*, 35 (3):1–17, January 2020. Cambridge University Press.

[26] Michael Slavik and Imad Mahgoub. Applying machine learning to the design of multi-hop broadcast protocols for VANET. In *Proceedings of the 7th International Wireless Communications and Mobile Computing Conference*. IEEE, July 2011.

[27] John Moore, Gerd Kortuem, Andrew Smith, Niaz Chowdhury, Jose Cavero, and Daniel Gooch. DevOps for the Urban IoT. In *Proceedings of the Second International Conference on IoT in Urban Space*, pages 78–81. ACM, May 2016.

[28] Manoharan Ramachandran, Niaz Chowdhury, Allan Third, John Domingue, Kevin Quick, and Michelle Bachler. Towards Complete Decentralised Verification of Data with Confidentiality: Different ways to connect Solid Pods and Blockchain. In *Proceedings of the Web Conference*, pages 645–649, Taipei, Taiwan, April 2020. ACM.

[29] Manoharan Ramachandran, Niaz Chowdhury, Allan Third, Zeeshan Jan, Chris Valentine, and John Domingue. A Framework for Handling Internet of Things Data with Confidentiality and Blockchain Support. In *Proceedings of the Extended Semantic Web Conference (ESWC)*, Heraklion, Greece, May 2020.